Teaching and Learning in the
(dis)Comfort Zone

D1224653

Teaching and Learning in the (dis)Comfort Zone

A Guide for New Teachers and Literacy Coaches

*Deborah Ann Jensen, Jennifer A. Tuten,
Yang Hu, and Deborah B. Eldridge*

Foreword by Sandra Wilde

TEACHING AND LEARNING IN THE (DIS)COMFORT ZONE
Copyright © Deborah Ann Jensen, Jennifer A. Tuten, Yang Hu, and
Deborah B. Eldridge, 2010.

All rights reserved.

First published in 2010 by
PALGRAVE MACMILLAN®
in the United States—a division of St. Martin's Press LLC,
175 Fifth Avenue, New York, NY 10010.

Where this book is distributed in the UK, Europe and the rest of the world,
this is by Palgrave Macmillan, a division of Macmillan Publishers Limited,
registered in England, company number 785998, of Houndmills,
Basingstoke, Hampshire RG21 6XS.

Palgrave Macmillan is the global academic imprint of the above companies
and has companies and representatives throughout the world.

Palgrave® and Macmillan® are registered trademarks in the United States,
the United Kingdom, Europe and other countries.

ISBN: 978–0–230–61769–8 (paperback)
ISBN: 978–0–230–61768–1 (hardcover)

Library of Congress Cataloging-in-Publication Data

    Teaching and learning in the (dis)comfort zone : a guide for new
teachers and literacy coaches / by Deborah Ann Jensen...[et al.].
      p. cm.
    Includes bibliographical references and index.
    ISBN 978-0-230-61769-8 (alk. paper)
    1. Language arts (Elementary)—United States. 2. First year
teachers—Training of—United States. I. Jensen, Deborah Ann.

LB1575.8.T425 2009
372.6′044—dc22   2009039979

A catalogue record of the book is available from the British Library.

Design by Newgen Imaging Systems (P) Ltd., Chennai, India.

First edition: January 2010

10 9 8 7 6 5 4 3 2 1

Transferred to Digital Printing in 2012

Contents

Foreword

Sandra Wilde

Wait, isn't life supposed to be about getting *more* comfortable? The authors of this exciting book have turned that conventional wisdom on its head. They demonstrate that in our work as teachers, the margins where comfort and discomfort overlap are not only the spaces where professional growth occurs but the key to staying engaged and happy throughout our careers.

There's a conventional wisdom about teaching careers: that there's no career ladder, that teachers leave the profession because every year is the same and there's only room for a few to get promoted to administrative positions. Visions of aging teachers using the same stale lessons and faded bulletin-board displays for thirty years.

Teaching is, however, a profession (the one devoted, after all, to the life of the mind) where reflection and growth must be central. Our knowledge and understanding of four commonplaces of education—teacher, learner, subject matter, and milieu (Schwab, 1977)—constantly change, and the good teacher grows as a professional in large part through continuing to respond and adapt through learning and reflection. We understand more than we used to about the role of the teacher, the diversity of learners, the intricacies of subject matter, and the role of education in serving all members of society. As professionals, we don't stick with what we learned in our initial training but keep up in the field, not only through the traditional methods of reading journals and attending conferences but through constantly examining and changing our practice.

Just as in medicine, practice alters as knowledge in the field grows and changes, and practitioners (physicians and teachers) are part of a community that develops in sophistication over time. The role of the literacy teacher has changed from someone who reads out of a teacher's manual to someone who understands how learning happens, how students differ from each other, what reading and writing involve, and how language and culture are intrinsic pieces of children's lives.

Yet it's not always easy for teachers to get to this point. This is where the (dis)comfort comes in. It's a challenge, for instance, to recognize that a lesson didn't go well, or that some of your students aren't improving in reading, or to invite a colleague to observe you and give feedback. This book takes lessons learned from working with teachers in a graduate literacy program and turns them into tools that teachers can use on their own to ramp up their practice. The authors' great insight is that the internal cognitive and emotional processes of professional growth can be jumpstarted by practical, empirical practices such as asking students to write exit notes about what they learned, or thinking about the language you use with kids, or examining how your teaching is addressing content as defined in state standards.

These avenues to growth take us out of our comfort zone, but not in a bad way, no more than getting up from our comfort zone on the couch to go hear a lecture is a bad idea. Just like kids, teachers benefit from working in their zone of proximal development (Vygotsky, 1978), pushing themselves beyond where they are at the present moment. And this is what makes teaching not only rewarding but also fun. Who needs a new job title when you can be developing lesson ideas based on your assessment of students, joining other teachers in study groups to explore what's new in the field, and getting to really know and understand your students' families and community?

One of the many things I love about this book is the way that it provides invitations to teachers at every level, from preschool to the university. Everyone's had those moments where a class just didn't go well and you start to question your whole philosophy of teaching. All teachers worry about the students who are harder to reach, who have a harder time with learning. All of us wonder if we're connecting with students whose cultures differ from our own. *Teaching and Learning in the (dis)Comfort Zone* invites us not to jump off the cliff but to dive into the pool, with tools we can use to swim rather than sink, and make our work lively, thoughtful, and satisfying rather than merely comfortable.

References

Schwab, J.J. (1977). Structure of the disciplines: Meanings and significances. In A.A. Bellack & H.M. Kliebard (eds), *Curriculum and evaluation* (pp. 189–207). Berkeley, CA: McCutchan.

Vygotsky, L. (1978). *Mind in society: The development of higher psychological processes*. Cambridge, MA: Harvard University Press.

Preface

We all love a good story. During the rare moments we have together over a cup of coffee or minutes before a meeting, we exchange stories of our lives: of our children, our families, our friends, the schools we have visited, and the sometimes funny, powerful, or woeful tales of our teachers. Many of these stories have informed our practice and helped to shape our curriculum in response to what we have been told and to what we see in schools. And we share these stories in our classrooms to help illustrate our own learning and teaching. Now we would like to share some of the stories with you.

A Community of Learners

We are from a learning community that can be characterized as a group of learners who "focus on problem solving and inquiry,... learn through action, reflection, and demonstration; and establish a learning atmosphere that is predictable yet full of real choices" (Short & Burke, 1991, p. 5). As faculty members in a graduate program in Literacy Education in a public college located in a large metropolitan area, we strive to embed these characteristics in our program, curriculum, and instruction. We bring to our teaching and research years of experience working in schools as teachers ourselves, staff developers, and curriculum specialists. We also bring our experiences as avid readers, productive writers, world travelers, and mothers of seven children ranging in age from nine to young adult.

The majority of our students are initially certified full-time teachers who work in the city public schools. Some work in charter, private, and parochial schools in the city and also in suburban areas. Most of them are in their twenties, with one–three years of teaching experience. They bring with them to our learning community their increasing understanding of children with diverse cultural, socioeconomic, and linguistic backgrounds and learning processes. They are trying to balance the complexities of teaching literacy in inner city schools while adapting to the challenges of the multidimensional demands of teaching and learning in one of the largest urban school districts. At the same time they are pursuing graduate degrees.

Our Program Orientation

Created in 2001 by a group of professors whose work in reading instruction was recognized as exemplary by the National Commission on Excellence in Elementary Teacher Preparation for Reading Instruction (Hoffman et al., 2005), our masters program in Literacy Education leads to a state professional certification in Literacy Education. Most of our graduates continue to teach in the classroom. Some have become literacy coaches, grade level leaders, mentor teachers, reading specialists, resource room teachers, and curriculum specialists. The program is designed to provide courses in content knowledge of reading and writing, in pedagogical skills of literacy instruction, in problem-solving skills required to assess learning difficulties and providing differentiated instruction, as well as leadership skills in literacy education. Evidence-based practice and fieldwork are integrated in every course to provide opportunities for teachers to learn while attending to the realities of classroom literacy curriculum and instruction in a culture of high-stakes testing.

About this Book

This book has grown out of our own teaching, learning, and mentoring experiences. It is a book about teachers learning to teach literacy, about the processes of teacher development from novice teachers to expert teachers of literacy, about mentoring teachers in reading and writing instruction, and about ways to maximize the learning potential of both teachers and their students. It is a book about the (dis)comforts we have found teachers experience on the road to becoming literacy leaders, to acknowledge them as opportunities to learn and grow, and to provide their stories as springboards for questions, challenges, and conversations. We invite the reader to explore and learn alongside our teachers and to consider the tensions and (dis)comfort in his or her own classroom and the possibilities they present for growth.

As we share our stories and those of the teachers with whom we work, we will often use the singular I instead of we in order to stay true to the lived experiences. We refer to our students as the teachers they are and not as students in order for the reader to differentiate our students from the children with whom our teachers work. The names of the teachers and children are pseudonyms.

About the Organization of this Book

The opening chapter of this text "Where Is Your (dis)Comfort Zone?" recognizes the need for new teachers to carve out a learning space in their often demanding and challenging induction years as teachers, and the crucial role of support for professional development during these years as new teachers teach and learn to teach well. This chapter illustrates what we mean by a (dis)comfort zone and why it is important for teachers to locate theirs. The book is then divided into two parts: Part I—Teaching and Tensions; and Part II—Tensions of Literacy Teaching.

Part I, "Teaching and Tensions," focuses on the tension or discomfort in teaching. This section is designed for the teacher to look inward at her own teaching, learning, and classroom language causing her to be reflective of the choices she makes for best practices. Chapter 2, "Choosing Words," deals with instruction and the language of the classroom, how we choose words when interacting with children, and how words help create the culture of the classroom. Chapter 3, "Cultivating a Reflective Stance," discusses why and how teacher reflection can foster professional growth. A variety of strategies and tools are provided to help teachers explore and establish supportive conditions in order for reflection to become an integral part of teaching and learning. Chapter 4, "Observing the Lesson," discusses the benefits of having the opportunity to view another teacher's classroom, of having your own teaching observed, and the rewards found in the reciprocal process of learning through feedback. Chapter 5, "Acting into New Ways of Thinking," promotes the idea of putting teachers in the shoes of learners. Stories, discussions, strategies, and tools in this chapter focus on helping teachers confront their own personal and professional knowledge, and through meaningful learning experiences, develop new perspectives into their teaching and their students' learning.

Part II, "Tensions of Literacy Teaching," more specifically deal with topics that cause discomfort in literacy education. It is designed for teachers to look outward toward their students and the families with whom they work. Chapter 6, "Understanding and Moving beyond Labels," explores the preconceptions teachers have about learners and their needs. With a focus on English Language Learners, struggling readers, and gifted children, the chapter helps teachers to separate judgments from observations as they work with a challenging student. Chapter 7, "Teaching Effectively Means Learning from Our Students," explores the question: What should I do when the planned lesson is not reaching the children? It speaks of the frustration teachers feel and the tension created in trying to follow a plan and the anxiety they feel when diverting from the plan fearful of leaving out important concepts and strategies. Chapter 8, "Learning from Parents and Families," emphasizes the importance of establishing a home-school connection around literacy practices to better ensure student success. It recognizes the tensions teachers experience and difficulties they feel when working with parents but equips the reader with strategies to establish partnerships. Chapter 9, "Teaching under the Accountability Umbrella," presents the tension that exists for classroom teachers as they try to implement good teaching practice in a culture of high-stakes testing and numerical measures of student achievement. It helps teachers reflect on their teaching of test content rather than teaching to the test. Chapter 10, "Pulling it

All Together beyond the (dis)Comfort Zone," focuses on teacher leadership and making one's own practice and decision making around literacy education and learning visible. What happens when your discomfort zone becomes your comfort zone?

Each chapter begins with an anecdote from our classes or classes of the teachers with whom we work to illustrate the topic being discussed. Following the anecdote we discuss the challenges and opportunities embodied or brought forth by the anecdote or problematize the often taken-for-granted notions regarding teaching and learning. We then present the research related to the chapter topic and discuss the chapter further as it is illuminated by or is illuminating the current understanding in the field of scholarship. We offer additional resources for teachers on the topic. Finally, we provide tools for practice, self-analysis, or self-reflection for the reader to use. We have field-tested the tools and found them useful in furthering teachers' experiences or in helping teachers be reflective.

We believe that learners' active participation is vital for their learning. We invite you, just as we invite our teachers, to participate as you read the book, in sharing the stories that might be familiar to you, in problematizing teaching learning, in strengthening your own theoretical understanding of why we do what we do, in reflection of the common challenges and problems we have witnessed in urban classroom, in experimenting with tools and methods for teaching, learning, and growing. Most of all, we would like to keep you company, providing a source of energy and support, as you strive to achieve professional development and continue in your growth as an educator.

We believe you will see yourself in the stories of our teachers and their schools. We hope they will give you comfort to know that you are not alone when teaching and learning in your (dis)comfort zone.

References

Hoffman, J.V., Roller, C., Maloch, B., Sailors, M., Duffy, G., & Beretvas, S.N. (2005). Teachers preparation to teach reading and their experiences and practices in the first year of teaching. *The Elementary School Journal, 105*(3), 268–287.

Short, K.C. & Burke, C. (1991). *Creating curriculum: Teachers and students as a community of learners*. Portsmouth, NH: Heinemann.

Acknowledgments

We want to thank our teachers for letting us into their classrooms, for sharing their stories with us, for challenging our own teaching and learning, and taking us to the edge of our comfort zones. For our teachers and their schools, this book is for you.

We are indebted to Camille Payne for her tireless effort in getting our work ready. We are thrilled to have worked with John Toth, a true art educator and artist. We would also like to thank the helpful comments we received from our reviewer, Charlene Klassen Endrizzi.

For my guys—Schelley and Isaak. Your continuous patience through this process, support, understanding, and love fills my heart to the moon and back.

Deborah Ann Jensen

I thank my mother, Carol Levine, for being my first literacy coach. And huge thanks to my family—John, Amy, and Maddy, for all their patience, humor, and love.

Jennifer A. Tuten

Thanks to my colleagues Deborah Jensen, Jenny Tuten, and Deborah Eldridge for your inspiration and support; to my daughters, Ocean and Azure, for your warm hugs and patience.

Yang Hu

To the teacher leaders in New Jersey who taught me the importance and nuances of their work.

Deborah B. Eldridge

Contributors

Deborah B. Eldridge is a professor and the dean of the Division of Education at Lehman College of the City University of New York. Prior to joining Lehman, Professor Eldridge was chairperson of the Department of Curriculum and Teaching at Montclair State University and earlier had been chairperson of the Department of Curriculum and Teaching at Hunter College. She is also a former assistant dean of education and NCATE (National Council on the Accreditation of Teacher Education) coordinator at Hunter. An expert in literacy, language arts, and teacher preparation programs, Professor Eldridge is the author of *Teacher Talk: Multicultural Lesson Plans for the Elementary Classroom* (Allyn and Bacon, 1998) and the forthcoming *Literacy for Life* (McGraw-Hill, 2010), as well as numerous other scholarly publications. She is an active member of several national professional societies, including the International Reading Association and the National Reading Conference.

Deborah Ann Jensen is an associate professor at Hunter College-CUNY primarily teaching in the graduate literacy program. She coordinates the two semester assessment and intervention program called Literacy Space where struggling readers and writers from New York City elementary schools work with graduate students. Her work with has been published in professional journals and presented internationally. As a literacy expert, Deborah also publishes and presents in the areas of parental involvement, using children's literature across the curriculum, and teacher education.

Jennifer A. Tuten is an assistant professor at Hunter College-CUNY primarily teaching in the graduate literacy program. She teaches courses in literacy assessment and remediation and practicum in literacy teaching. As project director of a professional development school partnership, she works extensively with elementary school teachers in East Harlem. Her research interests include literacy teacher development, teachers' understanding and experience of assessment, and parent/teacher communication. Her work on these topics has been published in professional journals. She has presented both nationally and internationally on areas of teacher development, struggling readers, and parent/teacher communication.

Yang Hu brings a rich and diverse background in teaching and staff development to her work as associate professor of literacy education and coordinator of the master's program in literacy at Hunter College-CUNY. Her Chinese/English bilingual and bicultural background led her naturally to her research interests in the literacy education of Asian immigrant students in American schools. She has published her work in this area in professional journals. Her roots in China beckon her to extend her research and teaching interest in China, investigating trends of the English writing instruction and sharing her expertise in literacy education with colleagues, teachers, and students.

Where Is Your (dis)Comfort Zone?

I was completely burnt out and felt as if I was powerless. All my ideals seemed to be fading as I realized how difficult my students were and how little my administration actually supported me. My principal was planning to retire and appeared to care very little about the success of our students. He interrupted my teaching once by calling me and asking if I was available to sing karaoke in his office with some other teachers. He never watched me teach a lesson in my classroom even ONCE. The general morale of the building was grim. When I began teaching two years ago, I was inspired, excited, and I wanted to change the world. Now my initial enthusiasm began to fade because I lost much of my positive outlook. I was not prepared for the issues that my students were dealing with. I know that I was doing great work at times, but I was not the teacher that I envisioned I would be. I began to lack energy to continue fighting. All of these finally led to my decision to move back home to work in my family's business.

—John, a former fifth grade teacher in a school located
in a low socioeconomic neighborhood

John knew teaching in a school located in the middle of a housing project would not be easy. He was aware of the high number of English Language Learners in the school as well as the number of children who qualified for free or reduced lunch. He was ready to embrace the diversity and the challenge. Yet the problems John faced in his school were none that he had anticipated. Not only did he have to face the challenges all new teachers encounter—to teach and learn to teach well simultaneously—he was also teaching in a school that even the substitute teachers would avoid. Disgusted with the lack of support, problems with students, and a general morose feeling across the school, he decided to leave the profession. We were deeply saddened by his decision. We knew that the city schools had lost a gifted, creative, knowledgeable teacher who taught from his heart.

Challenge and Opportunity

Teaching is one of the professions where there is no apprentice period. New teachers assume full responsibilities from day one. No teacher education program can fully prepare them for unanticipated demands and challenges in the schools. During the induction years, they must learn to juggle several balls—implementing the often unfamiliar curriculum; using strategies learned from their pre-service teacher preparation programs that they often believe to be the best practice; understand the divergent learning styles, levels, and needs of all students while trying to manage and organize the classroom, learn to communicate with parents, and try to recruit them as your partners; getting the students ready for state tests; and a myriad of unexpected problems that demand instant solutions on any given day. All this could stretch you to the edge of your comfort zone.

As the novice teachers say goodbye to their support network in the student teaching seminar, they often bring with them to their new teaching job an expectation that their schools would have a supportive environment where good teaching practices are shared and celebrated, and where a network of relationships would be formed to support their professional growth. The reality is that today's school culture is still one in which teachers work solo in their own classrooms with little support for new teachers to become effective in their work. The isolation has a deteriorating effect on the neophytes who can quickly become overwhelmed by the challenges and responsibilities. Their ideals shaken, they can begin to doubt their own abilities, the positive effect they could have on students, and their instructional decisions. Some question the decision they made to become a teacher; and many, like John, simply lay down their chalk to pick up new careers.

New teachers' needs and their school realities challenge our thinking about teacher learning. They also provide opportunities for us to problematize what has been taken for granted, and to explore and envision ways to mentor teachers as they journey through their induction into the teaching profession, which often include the most critical years of professional development. In our work as teacher educators, we know it is essential for us to support and sustain new teachers' learning so that they are invigorated and empowered during their induction years. We create learning communities that not only encourage them to solve problems they encounter in their schools, but also extend their learning in our community meaningfully into the realities and complexities of their classrooms and schools. In the literacy practicum course, we invite the teachers to identify areas of concern in their literacy instruction that is also challenging to them, and work with other participants, as well as the instructor of the course to learn and problem solve together. We took turns supervising the practicum course, where we felt the pulse rate of the school realities most strongly. We learned alongside the teachers, as we acted as their mentors, cheerleaders, and resource people. The genesis of this book came from the inspiring stories that teachers brought to our community.

Over the years, our work with teachers has provided us with ample evidence that the three pillars on which the Literacy Practicum was built have been essential elements for setting the stage for teacher learning and professional growth. In fact, these elements underlie much of our work in this book. They are:

1. A condition most conducive to learning is one that focuses more on action in problem-rich environments, interactive assessment, and reflection and less on judging and inspecting the learner's performance.
2. The learners need to be encouraged to work at the edge of their comfort zone (Daniels & Bizar, 1998). They need to focus on what they need/want to learn, instead of what they know they can do well. We encourage and support curious, questioning minds, prudent, professional risk-taking and self-challenge for growing as learners and teachers.
3. There is an emphasis on building a supportive as well as critical learning community. Members of the community act as a support team, collaborating to sustain one another's risk-taking and problem-solving efforts.

John's story is one among many that caused disequilibrium. Questions, challenges, and opportunities brought forth by his story, and other stories like his, urge us to frame and reframe our work and expand our understanding of teacher learning and growth in the context of real world classrooms. You will hear these stories throughout the book. In this chapter, you will learn that teacher learning often begins with identifying an edge of your comfort zone. You will learn how to recognize this zone, and how to use the discomfort as impetus for learning and professional growth.

Stories from the Field

Kathy was in a bind. She just survived her first year of teaching—a year filled with trial and error, the endless unexpected events that made many of her carefully planned lessons useless, one incident that involved a parent who threatened to sue the school, piles and piles of paperwork that kept her at the school until street lights were turned on, rushing to her graduate school classes twice a week after school and completing course projects on many a late night, just to name a few. As she entered her second year of teaching, she was switched to teach the third grade, a grade that would be taking the state tests in the spring. This meant that she had to learn how to implement test prep. On top of all this, her principal told her that for her observation in the spring, she wanted to see Kathy conducting writing conferences with her students. She was in a daze, feeling more frazzled than in her first year of teaching.

Kathy knew that she needed help. But she was hesitant to seek for it in her school. After all, just two months into her second year, she felt that looking for help was almost an admission of failure. But she had less than three months to prepare for her principal's observation of her conducting writing conferences. She turned to her graduate practicum seminar for help. Conversations with other teachers in the practicum as well as her practicum instructor convinced her to first examine and articulate what she already knew about writing conferences.

Kathy went into her first year of teaching with fourteen weeks of student teaching experience. Half of those fourteen weeks was spent in a fifth grade classroom where the cooperating teacher ran a writing workshop about three or four times a week. That was the first time Kathy experienced writing workshop. She loved what she saw in the writing workshop and felt confident about using the writing workshop when she began teaching. In the spring of her first year of teaching, she took the writing methods course and implemented the writing workshop at the same time. But she never felt confident about conducting writing conferences, especially after she learned that in the conference, the goal is to teach the writer, not fixing the writing. She thought that was always easier said than done. Besides that, she had a laundry list of what-ifs: What if the child writer refused to learn? What if she couldn't decide what the child writer should learn? What if the strategy she would teach the child wasn't what the child writer needed? What if the teaching point in the conference took away the child writer's ownership of his writing? What if the conference was too long? How could she keep other students on task while she conducted her conferences? Kathy felt overwhelmed, even resentful. The scale was seriously tipping toward one side—there was too little experience on the one side, yet there was too much to learn on the other. Her journey would be full of risk, but it was a risk that she had not choice but to take.

Unlike Kathy, Emily chose her own risk, not to fulfill any course requirements, nor to meet her principal's request. Emily teaches in a third grade Collaborative Team Teaching (CTT) classroom, where half of her students have been identified as special education students. Her co-teacher is a special education teacher with five years of teaching experience but is new to the school. Emily enjoys team teaching with her co-teacher partly because she finally can have collegial conversations all day, and partly because she feels that she has become more confident as a teacher after her first two years of teaching in this school.

Emily had wished that there were more support and mentoring for her at her school during those first two years. Her only conversations with the principal, besides the first one that was the interview and a recent one that led her to accept the CTT class, were the ones after the formal and informal observations. In those post-observation conferences, her principal pointed out to her twice that she needed to work on better organization and management of independent reading. Emily took the suggestions to heart. She reorganized her classroom library and children's independent book bins. She also revamped ways she conferenced with children and ways she kept her conference records. This year, the CTT class presented new challenges to the independent reading. Emily discussed with her co-teacher and decided that they needed to put their heads together and reorganize the independent reading to meet the needs of both the special education children and the rest of the class. She chose this not so much because it was in her discomfort zone, but because she needed to build on what she had learned in the first two years and make independent reading work for the CTT setting, which was itself a new challenge for her.

But soon Emily realized that independent reading had slid to the edge of her comfort zone. She tried giving students Post-its and asking them to jot down their questions as they read. She thought it was a great idea, as it would allow her to see evidence of student's engagement with reading. But it was chaotic for her students to manage the Post-its, and it was hard for many of them to understand the purpose of asking questions, some were not sure exactly where to put the Post-its, even after Emily and her co-teacher carefully modeled it and showed many examples. What was worse, Emily found that using the Post-its had slowed down children's reading, and in some cases, their understanding of the story. Emily thought that perhaps introducing Post-its was a total flop. She contemplated stopping the Post-its and wait until they found the right way for the CTT class to involve in independent reading. Should her experiment come to a sudden stop? Or should she push ahead and use this as a learning zone?

Research and Reflection

John is one of many who left the teaching profession. The National Commission on Teaching and America's Future (NCTAF, 2007) has indicated that the teacher dropout rate has grown by 50 percent in the past fifteen years, and that in the urban schools, it is over 20 percent compared to a 16.8 percent dropout rate nationally. They found that teacher turnover is highest in high-minority, high-poverty, and low-performing schools. They also reported that one-third of all novice teachers leave after three years, and 46 percent are gone within five years.

The financial cost has been reported to be approximately seven billion dollars a year (Kopkowski, 2008; NCTAF, 2007). And it is the high-poverty, low-performing schools that must spend their limited funding on teacher recruitment, hiring, and induction (NCTAF, 2007). Could you imagine what that amount of money put into our public schools could do?

But it is the students who suffer the most from having inexperienced and unseasoned teachers rotate through their buildings. "New teachers struggle, but as they gain more knowledge and experience they are able to raise student achievement. With the high rate of new teacher turnover, our education system is losing half of all teachers before they reach their peak effectiveness" (p. 4). Kopkowski (2008) makes the point that the schools least able to support new teachers are the ones where most new teachers are sent and where the more experienced teachers are needed.

Providing support to new teachers is essential in teacher retention. One of the goals of No Child Left Behind Act (2001) was to provide new teachers with induction programs where they would receive substantial support. Yet each school and/or district has interpreted support differently. Some provide a mentor who might meet with a novice teacher a few times a year while other schools may institute day-long workshops or classes. The new teacher does not necessarily receive regular support with an effective, veteran teacher leaving them to succeed or fare within the confines or their own classrooms (Ingersoll & Smith, 2004). The isolation they experience is vastly different from the collaborative and supportive student teaching environment and teacher preparation programs they just left (Kopkowski, 2008).

In low-performing, high-poverty schools like John's, highest rates of teacher turnover occur. Ingersoll and Smith (2004) report that "High rates of teacher turnover can inhibit the development and maintenance of a learning community. In turn, a lack of community in a school may have a negative effect on teacher retention, thus creating a vicious cycle" (p. 32). These schools are most frequently staffed with high concentrations of underprepared and inexperienced teachers, overwhelmed by the challenges and left to their own devices with little or no support or guidance (NCTAF, 2007). It has been established that mentoring programs are critical for beginning teachers (Darling-Hammond, 2003) and must come from a credible source, be informative, nonpunitive, and helpful (Mihans, 2008).

New teachers, as well as their support systems, must prioritize and understand the central tasks of learning to teach and focus on them. According to Feiman-Nemser (2001), the central tasks of learning to teach during induction years are "1. Learn the context—students, curriculum, school community. 2. Design responsive instructional program. 3. Create a classroom learning community. 4. Enact a beginning repertoire. 5. Develop a professional identity" (p. 1050). Feiman-Nemser also pointed out that there must be a commitment "to meet new teachers where they are and move their practice toward ambitious, standards-based teaching and learning" (p. 1027).

Even with the best support system in place, new teachers still need to learn how to identify their learning needs and create communities around their work. Learning theories that underlie effective instruction for children are also embedded in the framework in which adults learn. For example, the principle of gradual release of responsibilities (Pearson & Gallagher, 1983) can apply to teacher learning. However, this model implies that the teacher/learner must enter into a mentoring relationship, in which the teacher/learner is not only encouraged to learn, but also guided in practice, with feedback from the mentor, thus moving the teacher/learner forward in her development. Russian psychologist Lev Vygotsky (1978) used the term "zone of proximal development" to describe the learning process in which the learner is able to do more with the support of a mentor than she could on her own and extends her knowledge in the process. Some educators refer to this heightened learning experience as the "learning zone." In the case of teacher learning, we would like to believe that there is also a learning-to-teach zone.

For most beginning teachers, their learning-to-teach zone stretches most meaningfully from their student teaching to the first few years of teaching. Many have gained full control of basic content knowledge, pedagogical skills, and effective problem-solving skills during the first few years of teaching. As they enter a learning community, whether it be a graduate class, a school staff development workshop, or a school-based teacher study group, what becomes crucial for their professional development is that they must claim ownership of their own learning. Mallette et al. (2000) stated, "The real opportunities for learning came when students were asked to work in the field, make decisions about instruction, and have the time to interact with others to reflect and build knowledge and beliefs related to the area of literacy" (p. 594). Putnam and Borko (2000) indicated that when a diverse group of teachers come together in a community, they can draw upon and incorporate each other's expertise into their own teaching and learning situation.

More importantly, they must examine specific areas of instructional planning, teaching, communication with students and parents, as well as reflection in order to situate themselves in the learning-to-teach zone. This learning-to-teach zone must be content and context specific in order to develop contextualized thinking and learning, especially the complex pedagogical knowledge and problem-solving skills. The zone allows literacy teachers to sift through conflicting methodologies and practices, and to develop an individualized response to their students. It gives them the space for dialogue and reflection, to identify the tension points in their work with children, and build their knowledge and beliefs about literacy (Hermann & Sarracino, 1993; Malette et al., 2000; Roskos et al., 2000).

Is there a point along the comfort/discomfort continuum where teachers like Emily can be appropriately challenged? As figure 1.1 illustrates, the edge of one's comfort zone can overlap the edge of one's discomfort zone. We call this overlapping area (dis)comfort zone. We believe that this is the optimal entry point into a meaningful and transformative learning experience.

Locating the (dis)comfort zone is a crucial step, as the quality and nature of your learning process hinges upon knowing what you want and need to learn that is challenging yet manageable. You are best served when you

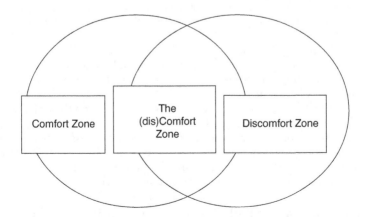

Figure 1.1 Locating your (dis)comfort zone.

are clear about your purpose, needs, and goals and when the ownership and responsibility for learning remain explicitly and authentically in your hands. You are ultimately the one who will do the learning.

Revisit the Story

"The course required me to learn at the edge of my comfort zone." wrote Kathy to her practicum instructor, "For me, I don't even have a choice. I was pushed to the edge of my comfort zone, well, more like over the edge."

"I am glad to see that you are willing to stay with the discomfort." Kathy's practicum instructor wrote back, "But perhaps you could explore further, are you on the edge of your comfort zone, or are you over the edge to the discomfort zone? Perhaps it would be helpful to distinguish and sort out what is within your control and what is not. Set some short term goals that focus on the edge of your comfort zone. Where do you start? What can you do before the winter break? What can be done after the winter break?"

The feedback from her practicum instructor helped Kathy to gain a new perspective on what she was facing. She came to the realization that learning to conduct effective conferences was on the edge of her comfort zone, whereas her principal's impending observation of her conducting writing conferences was over the edge. She now knew that she focused too much of her energy to cope with the frustration and anxiety of the observation, which was out of her control, that she lost sight of what was within her control—learning to confer.

Armed with the new insight, Kathy felt a load off her shoulder and set to work. She came up with a list of short-term goals and listened to suggestions. By the end of the fall semester, as she reflected on her work, Kathy was amazed to see that she had done the following:

- Establishing a system to keep conference records.
- Studying the conference records frequently to assess students' needs and progress.
- Taking students' writing notebooks home on weekends to become familiar with their writing, and to assess their needs and progress.
- Tape-recording her conferences, listening to the tapes to assess what worked well and what needed to change.
- Charting her strategy lesson ideas and posting them on the walls, so that she could refer to them during conferences.
- Focusing on building students' writing stamina, increasing their awareness of purposes and audiences for writing to keep their minds on their writing.
- Working on organization and management aspects of the writing workshop to minimize interruptions during the conference time.

More importantly, Kathy came to a new realization that what she was doing in the writing conferences included some of the best ways to differentiate the instruction. During her pre-service training, she came across this term many times. But it was until now that she felt deeply, for the first time, what it really meant to differentiate instruction. In the spring, Kathy was pleased to report that the principal's observation went well. In an email to her instructor, she wrote, "I think my children were so calm because I was calm...But more importantly, I took the risk and learned so much even though it was tough."

Emily, in her third year of teaching, never thought that she would learn in much the same way as her students. She admitted that allowing her students to put their own questions on Post-its in their books was a step away from her comfort zone. She had been used to being in charge of asking her students questions about books. The presence and support of her co-teacher sustained her. After several conversations with her co-teacher, Emily realized that the edge of her comfort zone was to invite her students to ask their own questions on Post-its. But this was a learning opportunity both for her and her students. Once they had questions on Post-its, Emily and her co-teacher could guide the students to ask better questions. This entry point led to a learning process that Emily owned and cared about.

The new insight led Emily and her co-teachers to focus on helping children ask better questions. They even developed a rubric to assess the range of questions questions. Their hard work paid off and the quality of students' questions improved. More importantly, Emily found that students were much more actively engaged in the reading process and they couldn't wait to share their questions. Then, she began to question herself about her own teaching. She asked, "Why was this [allowing students to ask question on Post-it notes] a major risk-taking for me? Did I not trust that they could ask good questions? Or was I afraid of losing control?" Later she reflected:

> my students [become]more engaged when they have opportunities to ask questions about the books they are reading. Like my students, when I ask myself questions about my teaching, I begin to see opportunities to grow as a teacher. For example, I realized that I could now teach them how to ask better questions. I guess I am going through the same process as my students.

Emily knows that she has become confident in her own teaching, but she is still growing in facilitating student learning. She is comfortable with asking her students questions about their reading. If her learning-to-teach zone is situated where she feels comfortable, her learning might take place along the lines of honing skills in asking better questions. (She could choose to do so if she feels that the main purposes for learning are to be judged or evaluated.) On the other hand, if she moves her learning-to-teach zone completely to an area where she feels uncomfortable losing control, she might focus on peer-led book clubs, in which her students would be encouraged to raise and answer their own questions. But this kind of learning could be an alienating experience because it is located in the learner's discomfort zone.

Strategies and Tools

Both Kathy and Emily felt rewarded by the outcome of their learning. A crucial contributing factor of their success was that they took on challenges of appropriate sizes and proportions, whether or not they chose the challenge or the challenge chose them. Another factor was that they didn't know what was appropriate and manageable without sizing up what they were facing. Only when they were grounded in their learning zone did they learn meaningfully, and feel empowered enough to face the fact that at the heart of their challenge was an anxiety about losing control in one-on-one conferences and inviting students' responses to reading.

Have you ever felt the cold sweats when you faced an overwhelming challenge? Like the time the teacher whose class you were observing asked you to help her with the running record, mistaking you for a student teacher? You were over the edge of your comfort zone into a discomfort zone. You would feel too anxious to take this opportunity to learn. The best learning zone is located at the edge of your comfort zone, or a (dis)comfort zone. Here you can expand your understanding, take on a new perspective or raise your awareness. The learning experience may be slightly uncomfortable, but you can manage your learning in incremental ways. How, then, do you find your (dis)comfort zone? You may try these strategies.

Self-Examination

You can conduct a rigorous self-assessment of what you need/want to learn or change. You can use the *Personal Inventory* in the Tools for Practice section at the end of this chapter to assess your need. You may need to prioritize your needs and wants. It is ideal if you can choose what risks to take. Perhaps you will come up with a challenge that you didn't choose to face. But be prepared if you are under pressure to meet certain requirement. Sometimes, external forces can push us into a (dis)comfort zone. You'll need to ask yourself some tough questions:

- Am I growing as a teacher?
- What do I need to do differently?
- Are my students learning? If not, what can I change to make that happen?
- Is there a part of my work that is challenging?
- Is there a part of my work that is perplexing?
- What doubts do I have about my work?
- Do I need to change my behavior first before I change my attitude?

The *Personal Inventory* at the end of this chapter contains specific areas of literacy education. There is a continuum of confidence level to describe each area of literacy teaching practice: *Confident/Effective, Somewhat*

Confident/Effective, Not Confident/Effective, and *Have Not Tried.* This inventory can help you identify your (dis)comfort zone. Areas you marked under *Confident/Effective* usually fall into your comfort zone. Also ask yourself, how do I know if I am confident or effective in this area? Do others in my school often ask me for suggestions in this area? Do I have evidence from student work that can proof my effectiveness? Did I get a satisfactory rating from my supervisor in this area? On the other hand, areas you marked under *Have Not Tried* may or may not be in your (dis)comfort zone, as you haven't tried them yet. We suggest that, after you complete the inventory, you examine areas that you've marked *Somewhat Confident/ Effective,* and *Not Confident/Effective,* and then identify one area that you want to focus on at this point in your development as a teacher, or an area that you need to focus on to maximize your student learning. This would serve as an entry point into a learning experience that is challenging as well as rewarding.

Feedback Examination

We can also get feedback from those around us, our colleagues, coaches, principals, students, and their parents, to determine where change or challenge is needed. The feedback can be in the form of informer conversations, post-observation conversations or evaluation letters, interview or survey of students, parent/teacher conferences. Quality feedback from these sources can help us determine what we need to learn.

Establish a Support Network

Both Kathy and Emily identified their (dis)comfort zones and faced their challenges because they both have a support system in place. Kathy used her practicum, and Emily's co-teacher was a co-learner. As new teachers, it is difficult to take learning risks without any support. When you are on the edge of your comfort zone, you need support and a sounding board— someone to help you problem solve when things don't go as planned, or someone to high-five you when things go well. Look for like-minded colleagues at your school, the teachers who teach the same grades, your literacy coach. If you are in graduate school, look for your peers or professors. You can improve the quality of your work if you organize communities around your learning.

Goal Setting

Once you identified the (dis)comfort zone, it is time to set goals that are focused in this zone, as well as explicitly address your learning needs. You may need to make the goals known to your support network, so that they can help you monitor and refine the goals. If you have too many goals, you'll need to prioritize them or focus on fewer critical ones. Goal setting will help you make your learning manageable and meaningful to you as well as to the students you teach.

The *Personal Inventory* at the end of this chapter will help you assess your professional knowledge and skills as a literacy teacher, You could be at different levels of development in different areas. This inventory can serve as a guide for self-reflection and assessment; it is a good place to start the conversation with your peers, mentors, coaches, administrators, or university teacher educators.

For new teachers, learning to teach is full of paradoxes. You need to demonstrate to your supervisor that you have the necessary skills and abilities to do a good job but you don't have them yet without sustained practice and learning. On the other hand, you can't hone in on these skills or abilities if you don't understand why you need them. But, every effective teacher must have experienced these paradoxes at the beginning of his or her career. If you stay on the path where your learning takes place at the edge of your comfort zone, little can hold you back from becoming the effective teacher you want to be.

Tool for Practice: *Personal Inventory—Finding Your (dis)Comfort Zone*

Please review your own practice/experience and evaluate your confidence or effectiveness in the areas of literacy instruction shown in table 1.1. There are blank spaces in each category for you to add your areas of concern. If you intend to check "confident/effective" for any area, ask yourself:

- How do I know if I am confident or effective in this area?
- Do others in my school often ask me for suggestions in this area?
- Do I have evidence from student work that can proof my effectiveness?
- Did I get a satisfactory rating from my supervisor in this area?

Table 1.1 Personal inventory

	Confident/ effective	Somewhat confident/ effective	Not confident/ effective	Have not tried
Creating and managing an effective learning environment that				
• fosters interest and growth in all aspects of literacy				
• has a library that supports students' independent reading and writing				
Planning curriculum and instruction including				
• establishing and articulating goals for student learning				
• taking into consideration students' needs, interests, and backgrounds				

Continued

Table 1.1 Continued

	Confident/ effective	Somewhat confident/ effective	Not confident/ effective	Have not tried
Facilitating and supporting all students in literacy learning, including				
• linking student learning to prior knowledge and experiences				
• encouraging critical thinking, inquiry, and problem-solving				
• demonstrating effective strategy use in reading and writing				
• conducting effective reading and writing conferences with individual students during independent reading and writing time				
• using small group instruction to scaffold student learning, such as guided reading and strategy lessons				
Understanding and organizing subject matter, including				
• using print and visual resources and technologies to make reading and writing accessible to all students				
• providing opportunities for learners to select from a variety of written materials, to read extended texts, and to read and write for many authentic purposes				
Assessing student learning to inform instruction				
• using a variety of assessment tools to collect students' learning processes and outcomes				
• using assessment data effectively to plan reading/writing instruction				
Developing as a professional				
• demonstrating a willingness to engage in collegial conversations with colleagues for continued professional growth				
• demonstrating a willingness to reflect upon your own practice in professional terms				
	Your Comfort Zone	Your (dis)Comfort Zone		Your (dis)Comfort Zone

Suggested Resources

Allington, R., & Johnston, P. (2002). *Reading to learn: Lessons from exemplary fourth-grade classrooms*. New York: The Guilford Press.

Braunger, J., & Lewis, J.P. (2001). *Building a knowledge base in reading* (2nd ed.). Urbana, Il: The National Council of Teachers of English.

Hoffman, J.V., Roller, C., Maloch, B., Sailors, M., Duffy, G., & Beretvas, S.N. (2005). Teachers preparation to teach reading and their experiences and practices in the first year of teaching. *The Elementary School Journal, 105*(3), 268–287.

Pressley, M., Allington, R. Wharton-McDonald, R., Block, C., & Morrow, L.M. (2001). *Learning to read: Lessons from exemplary first-grade classrooms*. New York: The Guilford Press.

Toll, C.A. (2005). *The literacy coach's survival guide*. Newark, DE: International Reading Association.

References

Daniels, H., & Bizar, M. (1998). *Methods that matter: Six structures for best practice classrooms*. York, ME: Stenhouse.

Darling-Hammond, L. (2003). Keeping good teachers: Why it matters and what leaders can do. *Educational Leadership, 60*(8), 6–13.

Feiman-Nemser, S. (2001). From preparation to practice: Designing a continuum to strengthen and sustain teaching. *Teachers College Record, 103*(6), 1033–1055.

Hermann, B.A., & Sarracino, J. (1993). Restructuring a pre-service literacy methods course: Dilemmas and lessons learned. *Journal of Teacher Education, 44*(2), 96–106.

Ingersoll, R.M., & Smith, T.M. (2004). Do teacher induction and mentoring matter? *NASSP Bulletin, 88*(638), 28–38.

Kopkowski, C. (2008). Why they leave. *NEA Today, 26*(7), 21–25.

Mallette, M.H., Kile, R.S., Smithe, M.M., McKinney, M., & Readence, J.E. (2000). Construction meaning about literacy difficulties: Preservice teachers beginning to think about pedagogy. *Teacher and Teacher Education, 16*, 593–612.

Mihans, R. (2008). Can teachers lead teachers? *Phi Delta Kappan, 89*(10), 762–765.

National Commission on Teaching and America's Future. (2007). *The high cost of teacher turnover*. Washington, DC: National Commission on Teaching and American's Future.

No Child Left Behind Act of 2001, Pub. L. No. 107–110, 115 Stat. 1425 (2002).

Pearson, P.D., & Gallagher, M.C. (1983). The instruction of reading comprehension. *Contemporary Educational Psychology, 8*, 317–344.

Putnam, R., & Borko, H. (2000). What do new views of knowledge and thinking have to say about research on teacher learning? *Educational Researcher, 29*(1), 4–15.

Roskos, K., Boehlen, S., & Walker, B.J. (2000). Learning the art of instructional conversation: The influence of self-assessment on teachers' instructional discourse in a reading clinic. *The Elementary School Journal, 100*, 229–253.

Vygotsky, L. (1978). *Mind in society: The development of higher psychological processes*. Cambridge, MA: Harvard University Press.

Part I

Teaching and Tensions

2

Choosing Words

Rachel, a fourth grade teacher at a large, sprawling suburban school, shared how she found herself paying attention to her praise of children, which was suggested by her professional book club discussion of *Choice Words* (2004) by Peter Johnston:

> In chapter four Johnston offers the suggestion of exclaiming, "I bet you're proud of yourself," and its implications, alongside of "I'm proud of you," and the implications of these words. I have actually been using the phrases, "I bet you're proud of yourself" and "how do you feel as a _____ today? (Reader, writer, researcher, learner, or whatever fits for the time), among others, and have found these words to be invaluable. My class has recently completed incredible science projects about natural disasters, and I found that the answers to these questions proved far more valuable than my saying, "I'm proud of you." The statement alone indicates that work should be done to impress or satisfy the teacher, when in fact students should be internalizing and feeling the rewards of hard work and achievement. Instead, I heard phrases such as "Wow, I can't believe how much work I did on this, it kind of feels really good," "I can't wait to show this to my mom," and "I like working this hard on something, it comes out so awesome at the end."

Rachel changed the wording of her praise to her students and it was a catalyst for her students' ownership of their growing academic identities. By reading and discussing *Choice Words* with her colleagues she is becoming more sensitive to her use of language in the classroom and its potential for promoting literacy growth.

Challenge and Opportunities

The distinction between "I'm proud of you" and "I bet you're proud of yourself" seems simple, even negligible. Being proud of a student for her good behavior or fine work is a natural feeling for teachers. The use of the word "I" in the statement directs the attention to the teacher. Altering the statement to "I bet *you're* proud of *yourself*," as Rachel did, shifts the focus to the student who can begin to own the accomplishment. And, as Rachel notes, students then can internalize that sense of pride in their work and strive for more success.

Language is the fuel of classrooms—it powers our instruction, our social interactions, and our assessments. Because it permeates all we do, we forget that our language has a powerful impact in our teaching. We take for granted that we make good word choices; indeed we frequently forget that we can make those choices in how we speak and write in the classroom. We are frequently unaware of what we say and how it is heard and understood by our listeners. We do not realize the powerful effect of altering our words, as Rachel discovered when she told her science students that they should be proud of themselves.

Becoming aware of our word choices and the impact of our choices is crucial to helping us support our students' literacy and overall academic, social, and emotional growth. As Johnston (2004) writes, "The language that teachers (and their students) use in the classroom is a big deal...Words and phrases exert considerable power over classroom conversations and thus over students' literate and intellectual development" (p. 10).

Teachers who have challenged themselves to become more conscious of their words through their reading and study of Johnston's *Choice Words* attest to the difficulties and rewards in paying attention to our teaching language. In this chapter you will share in their experiences, learn more about language and teaching, and explore several strategies to help you learn to choose words that promote literacy learning in your classroom.

Story from the Field

In his first year of teaching fifth grade, Andy Foster had his hands full. A new city, a new school, a new grade. While he felt right at home with teaching mathematics—keeping a brisk pace, relying on the certainty of equations and computations—he felt less comfortable in his literacy instruction, especially writing. His school had recently implemented a Writing Workshop approach but during his student teaching he had used a more scripted program. Now, with no script to follow, he found it difficult to manage and negotiate teaching children who moved through their writing at different paces. Some children would write very quickly, others seemed to spend days; and Andy was unsure how to keep it all together. During conference time, he felt he had to tell each student exactly what to do to improve the writing. "Make this a new paragraph" he told Max, crossing out words on his paper. "And I want you to add more details."

In January, Andy started a unit on children's books. He had his fifth graders study the picture books they had loved, look at illustrations, and begin writing their own books. The children were really engaged and he was feeling more confident. Then Mrs. Ballantine, the school principal, stopped by. The school's curriculum night, an opportunity for parents to learn about the work their children had done, was next week. She remarked, "These books would be wonderful to share with the parents. Make sure they are done and done well." Now Andy had the added pressure to ensure all the books were competed and that they "looked good." Some children were still revising, some were editing, some were working on their illustrations, while some were close to being done. "Class, I need you to get these books finished. This will be your homework assignment."

Grudgingly, the class complied with the requests, and the books were hurriedly completed in time for the curriculum night. On the night, parents seemed pleased with the engaging illustrations and charming stories. But Mrs. Ballantine, who only glanced at the covers, was unhappy with the overall appearance of the books and told Andy that she had expected more from him and his students.

Feeling somewhat frustrated, Andy graded the books. He wrote words like "sloppy, carelessly completed" on a number of rubrics, echoing the sentiments of his administrator. He had high expectations for his students and thought he needed to be critical of their work to hold them to a high standard.

As the students received their graded books, several were indignant. "Why did you say I was sloppy? You told me to rush!" complained Max. "I took your suggestions on the plot and now you say it is uneventful." Max pushed back his chair. Andy heard Max mutter under his breath, "I hate writing."

Andy hated hearing Max's statement. He wanted his students to like writing and become successful writers. What had interfered with this goal? How did Andy's language about his students' writing affect them? What impact did his choice of words in assessing Max's book have on Max?

Research and Reflection

Johnston's *Choice Words* served as the catalyst for turning our attention, along with our teachers, toward a deeper, more finely tuned understanding and appreciation of language. In this section, we first share the theoretical perspective that underpins this view of language. Then we present research that supports our understanding of the critical role teachers' language plays in creating a strong learning community and in fostering the literacy development of their students.

Language in the Classroom

As we stated earlier, language is the fuel or the medium through which teaching takes place. Cazden (1988) further asserts that in classrooms the teacher is responsible for controlling all the talk, a powerful position. It is also important to understand that spoken language is a part of the identity of all—teacher and students. Routman (2000) stated, "All learning involves conversation. The ongoing dialogue, internal and external, that occurs as we read, write, listen, compose, observe, refine, interpret, and analyze is how we learn" (p. xxxvi).

Language represents ideas, but also creates identities. It also positions people in relation to one another. Cazden (1988, p. 3) explains language functions in three ways: communication of propositional information, the establishment and maintenance of social relationship, and the expression of the speaker's identity and attitudes. All three functions, she argue, are intertwined. When Rachel says to her student, "You must be proud of yourself" after the science fair exhibit, she is providing information to him, sharing a relationship of mentorship with him, and fostering his own identity development as a science student as well as his sense of self-efficacy.

Research by Cazden and others (Heath, 1983) has shown that the most common pattern of classroom discourse is a three-part structure of teacher initiation, student response, teacher evaluation (IRE). It is the default pattern that teachers and students alike learn and understand as "school talk." As a familiar pattern it enables the participant to predict what is going to happen, a valuable factor. On the other hand, it can limit the kinds of language and learning opportunities. Teachers ask questions to which they already know the answer. In this way the teachers control what is to be covered in their classrooms. Classroom talk, then, is about checking for student understanding rather than for their construction of new ideas and insights. Students do not have the opportunity to deepen their understanding or engage in an intellectually challenging discussion of text.

Rowe (1986) investigated the importance of wait time in the classroom. She investigated talk in a variety of science classrooms and found that teachers typically waited a second or two before they called on someone or gave the answer. She found that when teachers wait three or more seconds, especially after a student's response, there are changes in the quality of the students' thinking and speaking. Hale and City (2006) suggest that teachers are too quickly apt to fill silence in a classroom by asking another question or sharing knowledge about a topic. They claim, "it allows students to be 'off the hook' in terms of generation their own thoughts and ideas about the topic. Students will often shift into cognitive neutral is they know the teacher will fill the silence" (p. 31).

Many teachers feel uncomfortable with pauses or silence in the classroom. Christina, a third grade teacher, had never considered giving students more time; she believed that if she wasn't talking she wasn't teaching. Encouraged to try, she found that waiting allowed students time to prepare what they wanted to say and modeled for her students that thinking can take time. She also found that with extending her wait time and supporting the thinking with phrases such as, "Let me think about this for a moment" quieter students, who did not usually volunteer, began to offer more comments in discussions.

Talking to Become Strong Readers and Writers

In a review of research about classroom discourse and reading comprehension, Nystrand (2006) explores the foundations for an understanding of the role classroom discourse plays in reading and writing development. The foundation was laid by Britton's (1969) notion of "talking to learn" and Barnes' (1976) distinction between transmission-oriented teachers and interpretation-oriented teachers. In a transmission-oriented approach, teachers view their role as providing students with knowledge, while interpretation-oriented teachers view their role as catalysts to guide students to new knowledge.

Vygostsky (1978) also informs this conceptual understanding in his framework. He says that teachers need to identify students' zone of proximal development and that students' cognitive development is enhanced by opportunities to talk, explain, and explore through dialogue. This "social formation of mind" (Wertsch, 1991) is a view that has led researchers in sociolinguistics (how language works in society) and psychology to investigate how and why discussion facilitates learning.

As important as it is to promote reading and writing in classrooms, it is also critical for teachers to pay attention to and help students develop oral language abilities. Indeed, particular kinds of classroom talk can promote academic success. Teachers model the language they want their students to develop. Over the past thirty years several different approaches have been developed that draw upon the belief that effective reading and writing development needs a scaffold by a collaborative approach to creating meaning. Reciprocal Teaching (Palinscar & Brown, 1984) was developed to foster comprehension of expository text by providing students with a structure within a small group to discuss a text. In this approach students take turns being the teacher and leading the group through four strategies: clarifying anything in the text that students don't understand, predicting information from the title and then from previous paragraphs, summarizing the paragraph, and creating a question that can be answered from the text. The teacher models each strategy with the goal of making students able to use and internalize the language and the strategy of clarifying, predicting, summarizing, and questioning, first in their own small groups and then independently as they read.

Questioning the Author (Beck et al., 1997) offers a different approach to modeling talk to facilitate reading comprehension. Rather than using specific strategy words, as in the Reciprocal Teaching model, teachers generate content-free questions that they call queries, "designed to assist students in grappling with text ideas as they construct meaning" (p. 23). The purpose of these queries is to help students attend to the author's intentions in the text. General examples include: What is the author trying to say? What is the author's message? What does the author mean? Does the author explain this clearly? More specific questions, about character, plot, setting, facts, and so on can then be developed as students discuss the text. A key component in this approach is that students participate in the discussion and collaborate in the construction of meaning, instead of the questions being used to share opinions or for the teacher to assess students' comprehension. The discussion takes place during the first reading of the text rather than afterward.

The teacher's role in the discussion regarding Questioning the Author is critical. Beck et al. describe several kinds of discussion moves that can provide models for students and provide openings in the discussion. "Marking" is a comment that draws attention to a student's remark, either by emphasizing a word in a paraphrase and adding emphasis, or explicitly noting its importance. "Turning back" is a discussion move that can return the discussion back to the students or by returning to a statement previously made in the discussion. This can help students learn to respond to each other rather than through the teacher. Turning back can also move the discussion back to the text. "Re-voicing" is yet another move. In this technique, a teacher can support a student's process of thinking aloud by restating what may have been expressed, with greater clarity. Modeling helps foster students' development by offering concrete examples of how to participate in the discussion. "Annotating" is a discussion move used by the teacher to fill in gaps in the discussion. "Recapping" is a move useful when the discussion has reached a point to summarize and move on. As teachers employ these moves in discussions of texts, their language provides a scaffold for students' comprehension.

Collaborative Reasoning (Chinn et al., 2001) is another model to support development of discussion-oriented literacy classrooms. The goal of Collaborative Reasoning is "to create a forum for children to listen to one another as they think out loud...to promote growth in students' abilities to engage in reasoned argumentation" (p. 183; Clark et al., 2003). In a Collaborative Reasoning discussion a small group of mixed-ability readers meet together. They all read a story together, one selected by the teacher to have rich possibilities for discussion. The teacher poses a question about a dilemma contained in the story. The students take turns explaining their ideas about the question. They are guided and urged to support their answers by using details from the story and connections from their own everyday experiences. Toward the end of the discussion the teacher polls the group and reviews the discussion.

Another approach to foster effective classroom discourse, now familiarly known as "accountable talk," grows from a constructivist view of language that argues that the social interaction with specific kinds of talk is essential to enabling students to develop deep understandings of complex concepts (Michaels et al., 2004).

Accountable talk, according to Michaels et al. (2002), has three components. The first facet is accountability to the learning community. This is talk that demonstrates attention to other participants, careful listening, and questions that aim to clarify or draw out ideas. When teachers begin to model such phrases as "Does anyone else want to add to this point?" or "Can you explain what you meant when you said—?" or "Take your time. We'll wait," students begin to use them as well. When the teacher regularly uses these kinds of expressions and questions, she or he establishes them as the norm of the classroom. This becomes the way talk occurs in the classroom, which then becomes embedded in the language of the learning community. It is also important to note that for this kind of conversation to truly take hold, the content of the discussion must be deep, relevant, and academically important.

The second component is talk that values looking for explanation, searching for ideas, and reasoning. This talk is the ability to participate in rigorous thinking. Some kind of questions that promote this are: "Why do you think that?" or "Can you say more?" Teachers help students become aware of the chain of reasoning through the threads of the discussion.

The final component in Accountable Talk is that of accountability to knowledge. As Michaels et al. (2007) state,

> Talk that is accountable to knowledge is based explicitly on facts, written texts or other publicly accessible information that all individuals can access. Speakers make an effort to get their facts right and make explicit the evidence behind their claims or explanations. They challenge each other when evidence is lacking or unavailable. (P. 289)

Teachers would press students to use the text to support their statements. Phrases such as "How do you know that?" or "Can you give me some examples?" or "I know that because it says here" would be part of this discussion.

Wolf et al. (2006) investigated the quality of classroom talk and academic rigor in reading comprehensions, looking specifically at aspects of Accountable Talk. By analyzing students' ratings and lesson transcripts, they found those facilitating students' opportunities for academic discussion and promoting development of skills led to academically complex and rigorous reading instruction.

To summarize, these models of classroom instruction share an understanding of the important role that purposefully modeled and led discussion plays in supporting the growth of literacy skills. They share an emphasis on authentic dialogue, teaching and modeling of interactions between teacher and student and with an overarching goal of supporting students in studying text and becoming more independent as readers and writers.

Language of Assessment

As well as using language in the classroom, teachers also write to assess children's work. Report cards, for example, embody teachers' written language and so the choices here have power and impact as well. Lomax (1996) interviewed pre-service teachers about their preparation for writing report cards. He found that despite completing an assessment course, they felt unprepared for the difficulties of grading and report card writing. Experienced teachers also find writing report cards challenging. In their study of the report card writing processes of language arts teachers,

Afflerbach and Johnston (1993) reported teachers' conflicts between what they knew and valued in their work with children and the system and structure of the report cards they were required to write. Concern was expressed at "the ability of the information required by the report cards to portray students' achievement to their satisfaction and to having their personal knowledge of students not valued" (p. 81). Additionally, Afflerbach and Johnston described a variety of purposes for report cares: for example, to report test scores, document behavior, or praise. There are also several audiences for the report cards, for example, parents, other teachers, administrators, and students. These factors contributed to the difficulties teachers experienced in writing report cards. They concluded, "We did not anticipate how stressful teachers would find the report-writing process or how severely the process would be affected by the constraints under which teachers worked" (p. 85).

Tuten (2007a) describes the dilemmas a novice teacher faces as she struggles to describe the students she teaches on her city's report card. Not only does she find the categories on the card limiting, but her school administrator carefully monitor the report cards and she is given a list of comments to use. Those comments, "Works hard" or "Needs to be consistent with homework," were too generic. The phrases on the report card, too, didn't reflect what she wanted to communicate to parents. She said,

> I felt that "Demonstrates effort and completes homework" especially because of the families in our school, I feel that that shouldn't have been coupled together because there are a lot of kids that I know and working with the families and working with them that a lot of them, for practical reasons have a lot of difficulty completing the homework at night and sometimes it doesn't necessarily mean that they're not demonstrating effort. (P. 36)

Parents care deeply about what is said in the report cards (Tuten, 2007b) and look to the report card language to learn more about their children. Lawrence-Lightfoot (2003) talks about the "essential conversation"—the dialogue between home and school—that can make both teachers and families very uncomfortable. It is thus important that teachers recognize the importance of considering the impact of their words in report cards and in parent-teacher conferences so that they are able build bridges with families as well as students.

Revisiting the Story

Andy shared his frustration with his Writing Workshop in his literacy practicum course. "I really want them to enjoy writing, but there is so much to do. I feel like I need to tell the students what do at each stage," he explained. Jane, a fourth grade teacher with several years of experience, seized upon his choice of the word "tell."

"Andy, you said *tell* not *teach*. I think there is a big difference between telling kids what to do and teaching them how to do it?"

Andy considered this distinction and thought about it. Perhaps in his desire to get his students to produce accomplished pieces, he had been telling them what he wanted, rather than teaching them to be writers. Telling it, as he thought about, did not enable the students to figure out anything on their own.

Jane reminded Andy of their discussion of Choice Words and suggested that he consider, purposely using one of the phrases to help him move from telling to teaching.

Over the next weeks, Andy made an effort to be more aware of the language he used in his conferences. He tried to pull back on telling his students what to do, and tried instead to let them lead. Andy still had Max's disappointed words ringing in his head and at the next conference, on Max's realistic fiction story about a boy who was obsessed with skate boards, he tried a different approach.

"Max," Andy began. " You really have me interested in Pete because of the things he says, and if you show me how he says them and what he looks like, I will get an even stronger sense of him."

Max leaned forward, now more eager to consider revising the story. Andy's comment drew attention to what Max had done well by showing the impact of the writing on the reader. But it also pointed out the next steps for Max to grow as a writer. Notice too that Andy said "and" instead of "but." The structure of what he said communicated to Max the potential the story had and Andy's belief that Max, as a writer, could revise the story to meet that expectation. This was a teaching move, rather than a critiquing or telling one.

Strategies and Tools

Reading *Choice Words* in professional study groups provoked much discussion and reflection among teachers. It pushed language to the forefront of their consciousness. Teachers began paying more attention to the language they used in their teaching and started to look at the impact that their language choices had on their students. Diane wrote in her blog:

> Everything we say and do can have such an impact on our students it is really a terrifying concept to consider. *Choice Words* really brought this to the forefront. I agree that teacher language can be very empowering. I think that it is important that our attitudes match our words. Students can sense your belief in them and genuine caring and concern for them. This is a powerful motivator in itself.

Sheila also found herself reevaluating her language:

> In chapter 8 Johnston makes a few connections between teaching and parenting. When I read this, to tell you the truth, I found myself saying, "I sure

hope not!!!" I am not a mother and I don't know what parenting is going to be like. But I am a teacher and I know that I catch myself way too many times being in a bad mood, speaking monotonously, not smiling...scary stuff!

In the following sections we explore the particular ways in which teachers have begun to pay attention to the words used in the classroom. One way to begin addressing these issues is by participating in a professional book study group as a way to deepen your knowledge of literacy instruction and to build a professional learning community. Invite several colleagues to join you to read and discuss the book together over coffee or lunch. The Tool for Practice *Professional Study Group Guide for Choice Words* at the end of the chapter provides some guiding questions and activities we've developed to get you started with your discussion of *Choice Words*.

Creating a Classroom Community

How we speak with children about our classroom and the work we do is critical in creating a strong classroom community. Johnston explores the difference between a teacher using "I" and "We" in talking about work. In using "I" the focus is all about the teacher, while "we" creates a shared sense of purpose and responsibility. Tina writes, "I like the reference it made to remind our students that it is 'we' in the classroom and it is 'our' classroom. Constantly, throughout the day, I have caught myself saying 'I' about my classroom but then I remind myself that it is 'we.' This has helped my second graders take more responsibility for their work and for their behavior."

Anna shares her view from a fourth grade classroom:

> The importance of the concept of "we" has really struck home with me. As an upper grade teacher, sometimes I need to establish myself as the authority figure so that students will not walk all over me. However it is important that students should feel invested in the classroom community and understand that are a part of setting behavior goals. Also, they will feel more confident to share if they thing they are sharing with us (the class) rather than me (the teacher who appears to be looking for a certain answer).

Giving Praise, Changing Behavior

We all want to acknowledge and affirm the successes of our students. Yet we rarely dig beneath the surface of how we praise. As we discussed how Johnson wrote about the banality of "Good job," many teachers took exception, stating that praising children in that way was effective. But with further discussion, debate, and a willingness to try a different type of praise, teachers found themselves moving toward offering students specific compliments designed to affirm and extend work, much as Andy learned to do in his conferences.

Patty writes, "As a teacher am very conscious of the words that I use with my students. I am working with the ways that I use positive reinforcement and working to be clear and not vague." Risa added, "I have been moving away from key phrases like good, great job, well done and giving more concrete examples such as 'Your writing shows specific examples that support your main idea.' This way my students better understand what they are getting praised for." Peter finds giving positive comments such as "I love the way Tina supported her argument with a quotation" and "I notice how Artie is sitting quietly" effective.

As well as giving praise, carefully chosen words are critical, teachers discovered, when they need to reprimand. Upon her initial reading of the book, Mary Ann shared with group that she realized that she had become a yelling teacher, not the kind of teacher she aspired to become. She decided to make a concerted effort to use the phrase "That's not like you" instead of scolding. Doing so, she found, helped her students take ownership of their behavior and raised the expectations.

We are not always aware of the language we use in the classroom and it can be uncomfortable to acknowledge that we may be not communicating our best intentions to our students. Mary Ann's recognition that she had become a "yeller" and not the teacher she had aspired to be echoes comments of other teachers. Some realized that they inadvertently used sarcasm with students, which was ineffective and disrespectful to their students.

Creating Opportunities for Deeper Learning

As explained earlier, much classroom discourse consists of teacher question/student response, in which the teacher already knows the answer to the questions. Yet if teachers ask different kinds of questions, ones that authentically engage children in the learning process, a different kind of learning can occur. Instead of accepting correct or planned answers or rejecting "incorrect" ones, teachers can help students think about their learning, help them become more metacognitive by asking probing questions. Gina wrote:

Another idea presented that I enjoy is the question "What problems did you come across today?" I think this question really sets an open stage that not only allows, but welcomes obstacles and challenges. I think it also, when followed with a discussion, it can allow students to see that adults make mistakes too, and face challenges everyday. The question, what problems did you come across, can and should be used in many different conversations. In small moments, perhaps, in relation to a homework assignment or during independent reading, as well as in much larger moments, perhaps during an experience where conflict-resolution was implemented, or during a moment where a student was faced with a challenge in their personal life that they had never before experienced.

Another question, "How did you figure that out?" communicates to students a teacher's interest and affirmation of students as problem solvers. Responses to this question provide teachers with insights into students' thinking processes.

Thinking about how critical her role was in using language to support problem-solving and critical thinking, Ellen, a fifth grade teacher, wrote:

> When I began teaching I was always looking for the right answer. However now I believe in classroom there are so many right answers. In the book it mentioned the comment, "That's a very interesting way of looking at it. I hadn't thought about it that way. I'll have to think about it." Not only did I use it in my classroom but I now use it in my literacy block. When they think that they have explained something in a way that you didn't see it, they are very excited and confident.

To further expand her students' opportunities for discussion, she decided to use the principles of accountable talk as a framework to make explicit the kind of talk and learning she expected from her students. Ellen began with talk to build community as she introduced literature circles to her class. She took aside a group of students to coach in participating in literature circles. She shared with them some conversation openers or extenders such as I agree with ____ because, I want to add something, Can you explain what you meant when you said, and modeled them with her small group. Then she used the fish bowl technique to share with her whole class the kind of book discussion she wanted them to share in. After several weeks with the literature circles she reported, " I am amazed at how smart they are! They are digging into the reading and supporting their ideas with evidence!"

Reflecting on Words

As Johnston reminds us, teaching and our language in the classroom is to a great extent automatic. In this chapter we've suggested that you take a step outside of your comfort zone to think about the potential impact of what you say and to consider alternatives that may help you achieve even more in your instruction.

The two final Tools for Practice at the end of this chapter are designed to help you. *Bookmarks for Choosing Words* will help you try out new ways of speaking with your students with the aim of supporting their literacy abilities and the development of literate identities. On these bookmarks are phrases and sentence openings that can help you experiment with changing the way you speak in class. You can clip them to your lesson plan or conference notebook.

The Tool for Practice *Listening to Myself* invites you to tape record yourself teaching so that you can begin to be aware of the words you use, the tone of your voice, and the conversational patterns in your classroom. Listening to yourself is the first step to help you look critically at the language interactions between you and your students so that you can identify any patterns and make changes.

Tool for Practice: *Professional Study Group Guide for Choice Words*

Reading and discussing a book with a group of colleagues is a great way to share ideas, keep current with professional literature, and continue to be an active reader (a role you want to model for your students). We've created these questions to get you started.

1. Johnston begins chapter one of *Choice Words* with an anecdote about a teacher who used playful and imaginative language as a way to change his student's misbehavior. Think back to the teachers you have had. Who stands out in their use of language? Why?

2. "Did anyone notice?" is a question that invites students to take an active role in learning. Think about your literacy lessons. When, where, and how might you use this question in your teaching?

3. In our classrooms, Johnston (2004) writes, "children are *becoming* literate. They are not simply learning the skills of literacy. They are developing personal and social identities—uniquenesses and affiliations that define the people they see themselves becoming" (p. 22). How do you understand the relationship between teaching the skills of literacy and supporting students' development of literate identities?

4. Agency, according to Johnston, is a sense that one can act strategically and accomplish their goals. As you read and discuss this chapter, keep in mind the students in your class who seem to have a strong sense of agency and those who doubt themselves. With your colleagues, brainstorm a list of characteristics for each group. Then discuss what strategies you can incorporate in your teaching to ensure all children have an opportunity to develop a strong sense of agency.

5. In this chapter, Johnston analyzes the uses of several short phrases—"How else?," "That's like," and "What if?" These phrases, he argues, encourages flexible thinking and new ideas. Try to use these phrases with your colleagues as you discuss the book or at a grade level meeting. See if using these phrases pushes your discussion to a deeper level.

6. In the book (pages 62 and 63), Johnston suggests several activities to help alter the conversation patterns in your classroom. Have each member of your group try a different technique and report back to the group on what you each discovered.

7. "Are there any other ways to think about this? Any other opinions?" It can be difficult to entertain real difference of opinion in the classroom. How do you feel about this?

8. Now that you've read and discussed the book, what are three phrases or ideas that stand out as most relevant to your teaching? Why? What steps would you like to take next?

Did anyone notice?

What are you noticing? Are there any patterns?

That's not like you.

What a talented poet you are.

I wonder, as writer, if you are ready for this.

I bet you are proud of yourself.

What are you doing as a writer today?

What have you learned most recently as a reader?

How did you figure that out?

What problems did you come across?

Where are you going with this piece of writing?

How else . . .?

That's like . . .

What if . . .?

That's an interesting way of looking at it. I hadn't thought of it that way.

How could we check?

I wonder . . .

What are you thinking? Stop and talk to a neighbor.

Figure 2.1   Bookmarks for choosing words.

Tool for Practice: *Bookmarks for Choosing Words*

Many of the teachers who read and studied *Choice Words* wanted concrete reminders of words and phrases discussed in the book. Sara explained, "We make so many decisions a day as teachers, it is easy to overlook the choice of words we make when interacting with every student, I would like to make a list of the prompts that were used and attach them to my conferring clipboard and carry them with me during face to face interactions."

These bookmarks (figure 2.1) can be cut out and taped to your lesson plan or conferencing clipboard. Not a script, but a reminder of the kinds of words you might like to choose.

Tool for Practice: *Listening to Myself*

It is difficult to analyze discussions as they occur, but you will learn a great deal about your language choices by tape recording a lesson or conference. Teachers who have carefully studied their language are able to identify patterns, reevaluate word choices, and develop plans to more purposely use language in their classrooms. Once you become comfortable using this for your own language, you can consider sharing tape recordings of students' small group discussions or large group as an instructional tool to help students understand and develop their discussion skills. Here are steps to analyze a lesson.

1. Decide upon when and what you would like to record. A conference is often a good place to begin as there are only two participants.
2. Test out your tape recorder to ensure capturing the voices.
3. Tape your lesson.
4. Find a time and place to listen to your lesson—in the car on your way home, on your iPod.
5. Listen for your tone, the amount of wait time, the number of times you say "I" versus "we".
6. What patterns do you notice?
7. What would you like to change?

Suggested Resources

Beck, I., McKeown, M.G., Hamilton, R.L., & Kucan, L. (1997). *Questioning the author: An approach for enhancing student engagement with text.* Newark, DE: International Reading Association.

Denton, P. (2007). *The power of our words: Teacher language that helps children learn.* York, ME: Stenhouse.

Hale, M.S., & City, E.A. (2006). *The teacher's guide to leading student-centered discussions: Talking about texts in the classroom.* Thousand Oaks, CA: Corwin Press.

Johnston, P.H. (2004). *Choice words: How language affects children's learning.* York, ME: Stenhouse.

Lawrence-Lightfoot, S. (2003). *The essential conversation: What parents and teachers can learn from each other.* New York: Ballantine.

Nichols, M. (2006). *Comprehension through conversation: The power of purposeful talkin the reading workshop.* Portsmouth, NH: Heinemann.

———. (2008). *Talking about text: Guiding students to increase comprehension through purposeful talk.* Huntington Beach, CA: Shell.

References

Afflerbach, P.P., & Johnston, P.H. (1993). Writing language arts report cards: Eleven teachers' conflicts of knowing and communicating. *Elementary School Journal,* 94(1), 73–86.

Barnes, D. (1976). *From communication to curriculum.* Harmondsworth, UK: Penguin.

Beck, I., McKeown, M.G., Hamilton, R.L., & Kucan, L. (1997). *Questioning the author: An approach for enhancing student engagement with text.* Newark, DE: International Reading Association.

Britton, J. (1969). Talking to learn. In D. Barnes, J. Britton, & H. Rosen (eds), *Language, the learner, and the school* (pp. 79–115). Harmondsworth, UK: Penguin.

Cazden, C.B. (1988). *Classroom discourse: The language of teaching and learning.* Portsmouth, NH: Heinemann.

Chinn, C., Anderson, R.C., & Waggoner, M. (2001). Patterns of discourse during two kinds of literature discussion. *Reading Research Quarterly, 36,* 378–411.

Clark, A., Anderson, R.C., Kuo, L., Kim, I., Archodidou, A., & Nguyen-Jahiel, K. (2003). Collaborative reasoning: Expanding ways for children to talk and think in school. *Educational Psychology Review, 15*(2), 181–198.

Hale, M.S., & City, E.A. (2006). *The teacher's guide to leading student-centered discussions: Talking about texts in the classroom.* Thousand Oaks, CA: Corwin Press.

Heath, S.B. (1983). *Ways with words: Language, life, and work in communities and classrooms.* Cambridge, MA: Cambridge.

Johnston, P.H. (2004). *Choice words: How our language affects children's learning.* York, ME: Stenhouse.

Lawrence-Lightfoot, S. (2003). *The essential conversation: What parents and teachers can learn from each other.* New York: Ballantine.

Lomax, R. (1996). On becoming assessment literate: An initial look at preservice teachers' beliefs and practices. *Teacher Educator, 31*(2), 292–303.

Michaels, S., O'Connor, C., Hall, M., & Resnick, L. (2002). *Accountable talk: Classroom conversation that works* (CD-ROM set). Pittsburgh: University of Pittsburgh.

Michaels, S., O'Connor, C., & Resnick, L. (2007). Deliberative discourse idealized and realized: Accountable talk in the classroom and in civic life. *Studies in Philosophy Education, 27*, 283–297.

Michaels, S., Sohmer, R.E., & O'Connor, M.C. (2004). Classroom discourse. In H. Ammon, N. Dittmar, K. Mattheier, & P. Trudgill (eds), *Sociolinguistics: An international handbook of the science of language and society* (2nd ed., pp. 2351–2366). New York: Walter de Gruyter.

Nystrand, M. (2006). Research on the role of classroom discourse as it affects reading comprehension. *Research in the Teaching of English, 40*, 392–412.

Palinscar, A.S., & Brown, A.L. (1984). Reciprocal teaching of comprehension-fostering and comprehension monitoring activities. *Cognition and Instruction, I*, 117–175.

Routman, R. (2000). *Conversations: Strategies for teaching, learning, and evaluating.* Portsmouth, NH: Heinemann.

Rowe, M.B. (1986). Wait time: Slowing down may be a way of speeding up! *Journal of Teacher Education, 37*, 43–50.

Tuten, J. (2007a). "Between a rock and a hard place": A second year teacher's experience writing report cards. *Excelsior: Leadership in Teaching and Learning, 2*(1), 31–46.

———. (2007b). "There are two sides to every story": Parents negotiate report card discourse. *Language Arts, 84*(4), 314–324.

Vygotsky, L.S. (1978). *Mind in society.* Cambridge, MA: Harvard University Press.

Wertsch, J. (1991). *Mind as action.* New York: Oxford University Press.

Wolf, M.K., Crosson, A.C., & Resnick, L.B. (2006). Accountable talk in reading instruction. Educational Research and Development Centers Program. CA: The Regents of California.

3

Cultivating a Reflective Stance

Jessie (an assistant teacher): I am not sure if this is what we are supposed to do to reflect on our work.

Chloe (a second grade teacher): Let's see, what do we do when we reflect on a novel? We can't just retell what happened in the novel. We can't just comment on the character's action. None of these is reflection.

Jessie: That was pretty much what I did. I just added blurbs to explain what each item is in my portfolio. At times I made comments on why a particular item made me proud. I am not sure what else to say.

Chloe: When we reflect on a novel, don't we reveal how the novel has changed us, or what thoughts or feelings we have gained as we finish the novel?

Challenge and Opportunity

This was a conversation between two new teachers as they engaged in a peer conference about their own portfolios under construction. Anyone listening in to this conversation would probably agree that these teachers were involved in reflective thinking. Faced with a moment of uncertainty, they searched for answers that would settle their doubt. Chloe's analogy about reflection to novels points to one of the desired outcomes of reflection—that it leads to transformation or change. But if the target of our reflection is our teaching practice, not a novel, does reflection always lead to transformative outcomes?

Reflection has come to be widely recognized as an integral element in the teacher education programs as well as in teachers' professional growth. Terms such as "reflective practitioner" and "reflective teaching" come up frequently in discussions of classroom practice and professional development. In working with teachers in our Literacy Program over the years, we've seen teachers engage in a range of reflective behavior, from those similar to Jessie who struggled to find something to say to fill the reflection portion of their work, to those like Chloe, who contemplated on how reflection could be transformative. The challenge is how to help the teachers benefit from reflection so that they could be more sure-footed in their work as teachers. How can we use reflective lenses that will allow teachers not only to view their own practice but also themselves as learners? How do we cultivate a reflective stance that is crucial in the teacher's growth?

For early career teachers such as Jessie and Chloe, the first years of teaching and learning to teach are filled with uncertainties and perplexities. We recognize the importance of creating structures and using models of best practice in our graduate program to establish appropriate and supportive conditions to foster teacher reflection for both early career teachers as well as seasoned teachers. For the methods courses as well as practicum courses, reflection is hardwired into them. One of the opportunities to reflect is the web-based course management system. For example, at the beginning of these courses, we invite the teachers to reflect on their past experience and histories as readers, writers, and/or learners involved in various literacy events, as well as on their beliefs, assumptions, or perceptions. Then at the end of the course, we ask them to go back and reread their discussions posted at the beginning of the semester and examine their beliefs, assumptions, or perceptions in light of their learning throughout the course.

Besides the web-based discussions, teachers are also asked to reflect on their experience after trying out various instructional approaches and strategies. Here the teachers evaluate their experience, describe their students' learning, and look for ways to modify and adapt for the future.

As described in chapter four, the practicum course offers multiple opportunities for the teachers to be in charge of their learning as well as reflection: when they write to invite the instructor's visit; when they participate in the post-visit conversation with the instructor; when they write a letter reflecting on the learning that took place before, during, and after the visit; when they have conversations about their own and their partner's videos; as well as the web-based discussions.

Chapter five cites examples from an art-based portfolio construction. Here the reflection focuses on teachers' own learning processes as teachers learn through the arts. Another portfolio that the teachers in our program complete is their exit portfolio, which documents and describes their learning, growth, and reflection throughout their graduate studies. In this portfolio, they revise their philosophy of literacy education, which they wrote at midpoint in the program. This revision provides them with an opportunity to revisit their thinking about literacy a year or so ago. This portfolio also contains three artifacts of their own choices. These could be course work, fieldwork, or their children's work that show the impact of their teaching. For each artifact, the teacher attaches a page, explaining the context of the artifact, as well as reflecting on how the particular artifact shows growth or change in their learning.

Without a doubt, reflection is intrinsically bound up in teaching, as illustrated in the stories from the field in this chapter. We hope that this chapter will lead you to consider why and how reflection can foster professional growth, and understand that reflection can and should take place whether or not you are in a graduate program or working with a mentor. We encourage you to consider how you might explore and establish appropriate and supportive conditions so that reflection becomes an integral part of your teaching.

Stories from the Field

Celia teaches in a school situated two blocks from a large housing project. Most of her first graders struggle as readers and writers. Everyday, Celia goes into the classroom with an awareness that she should model what readers and writers do; and she should always use children's books to enrich her reading and writing workshop. During her first year of teaching, she made sure that when she modeled narrative writing in her writing workshop, she always included a beginning, a middle, and an end. After reading a realistic fiction book, she would always ask the children, "What does this story remind you of?" But to her disappointment, at the end of the year, most of her students were still writing one sentence to illustrate a picture, such as "I went to the store with my mom," or "I painted eggs for Easter." A handful of the children could write longer narratives containing a beginning, a middle, and an end. But they couldn't have done so without Celia sitting by their side, encouraging and helping them to write down their second and third sentences. Celia knew that she needed to do more than just modeling and using children's books to inspire their writing.

She began with goal setting in the fall of her second year of teaching. By aiming at building stamina first, Celia wanted to see her students become stronger writers who could push their writing from one sentence to at least three. "Three sentences," she said, "Nothing could be simpler than that for my students to understand." Besides modeling and immersing students in quality children's books, she also gave her students many opportunities to share their pictures and talk about their stories. After a few weeks of trying, she was disheartened to find that those with whom she worked during one-on-one conferences could write more than one sentences with her assistance, whereas the rest of the class still produced one sentence to illustrate their colorful pictures. Some were still content in just labeling their stories. What would she do?

More than one hundred blocks downtown, third grade teacher Andrea teaches in a school predominately serving children of Asian immigrants. Andrea has enjoyed her two years working in this school where she felt that parents really put a lot of faith and effort in their children's education.

In a social studies theme unit of the Native American Studies, Andrea infused many literacy activities into the various lessons. One of them was for her students to write creation stories under the influence of Native American creation stories they had read. But the students' writing "didn't go the way I had planned," according to Andrea. So she reread the two Native American creation stories, dissected them by making a web to illustrate the elements of the stories, and even created a diagram of four boxes, hoping that this would make it easier for her students to write. The children dutifully filled the boxes with ideas. But most of their stories sounded formulaic and dull, "almost like science reports," according to Andrea. Some of her students were confused and frustrated.

Celia's and Andrea's stories remind us that teacher reflection is rooted in watching one's teaching go off track. They view their students' difficulties as signals that their instructional planning and teaching didn't reach the intended goal. As teachers, most of us know that if our students didn't learn something as we had hoped or planned, then we need to modify our approach. The question that still needs to be asked is: How can we learn from our imperfections so that not only do we acquire more effective tools and strategies, we also are more effective at responding to our students' needs? More importantly, how can we cultivate a reflective stance so that we always keep an eye looking inward to tap into who we are, and how we are learning or changing as teachers and learners? How can our reflection go beyond subjective reasoning, referring only to our own personal history or experience to help us make sense of the new information?

Research and Reflection

In an informal survey, we asked teachers in our program three questions: (1) *What does reflection mean to you?* (2) *What do you do when you reflect on your teaching?* (3) *How does reflection affect your teaching and learning?* Three kinds of answers prevail:

- reflection is introspective;
- reflection is retrospective;
- reflection involves self-assessment.

Responses to the survey questions also seem to take it for granted that reflection is a desirable aspect of teaching. Many mentioned that reflection allows them to better meet the learning needs of their students. But very few explicitly point out that reflection can lead to transformation of one's own internal perceptions, beliefs, views, and capacities to learn.

Current research in teacher reflection regards John Dewey's work (1933) as fundamental to understanding the nature of reflection. According to Dewey, reflective thinking begins with "a state of doubt, hesitation, perplexity, or mental difficulty, in which thinking originates" (p. 16). The state of doubt could lead to a search for resources that will address the doubt or yield solutions. Dewey believed that thinking was natural, but reflective habit of mind needed to be taught. He claimed that the benefit of reflective thinking is a better control over the experience, therefore making it more valuable.

A more recent fundamental understanding of the nature of reflective practice was provided by Donald Schon (1983, 1987). Expanding on Dewey's idea, Schon believed that reflection is integrated in action. He delineated reflection as "reflection-on-action" and "reflection-in-action." Reflection-on-action suggests that teachers should learn to frame and reframe the complex realities and problem in their work, try out various interpretations and solutions, then modify their action as a result. Reflection-in-action assumes that reflection can take place in the process of acting, and simultaneously making modifications to the action.

The impact of the early work on reflection is seen in the understanding that reflection is an important process for teachers (Darling-Hammond & Snyder, 2000; Valli, 1997; Zeichner, 1996). An important feature of teacher reflection is that it is ongoing and cyclical (Clark, 1995; Korthagen, 1999; Loughran, 2002; Reiman, 1999; Stanley, 1998). Another feature of reflection is that broadening one's perspectives to include multiple perspectives is necessary to gain new insights into one's teaching and learning process (Collier, 1999; Hatton & Smith, 1995; Loughran, 2002; Rearick & Feldman, 1998).

Building on Dewey and Schon's theories, recent studies of teacher reflection have also described reflection as active and purposeful. In a critical review of empirical research on pre-service teacher reflection, Risko and her coauthors (2002) delineated that reflection is aimed at as problem solving, is regarded as a cognitive activity as well as a sociopolitical act, and is multifaceted. In addition, there has been much work aimed at describing developmental or hierarchical qualities of reflection. Van Manen (1977) put forth a framework containing three levels of reflection:

- Technical accuracy level: concerned with procedural accuracy.
- Reasoning level: concerned with rationale for instructional action.
- Critiquing level: concerned with how teaching can foster equitable conditions for learning.

Similarly, Valli (1997) created a hierarchy of five different levels of reflection as follows:

- Technical reflection: focusing on providing technical rationale.
- Reflection in/on action: focusing on pedagogical activity in the context of teaching.
- Deliberative reflection: focusing on considering other perspectives.
- Personalistic reflection: focusing on personal and professional growth.
- Critical reflection: focusing on weighing social and political implications of teaching and schooling.

Finally, Ward and McCotter (2004) describe four distinct levels of reflection based on their studies of reflection of pre-service teachers.

- Routine reflection: focusing on problem description; the self is often disengaged from change.

- Technical reflection: focusing on problem-solving in specific teaching tasks.
- Dialogic reflection: focusing on learning process, often considering students' and others' perspectives.
- Critical reflection: focusing on personal involvement in questioning fundamental assumptions in teaching, learning, and schooling.

The general principles about reflection as well as delineation of varying levels of reflection described in these frameworks help to demystify the reflection process. Knowing the functions of reflection, as well as differing degrees of reflection can help teachers resist the tendency to ask simply "How did I implement this curriculum?" or "Did my students get it?" Additionally, one must make teacher reflection a companion process to teaching, and strive to improve the qualities of reflection.

Now, let's go back to Celia's and Andrea's stories and see how they reflected and if their reflections helped them solve the problems they encountered in their teaching.

Revisit the Story

For Celia, her reflection was made visible to herself as well as to others because she was a participant in the practicum course. The milieu of the practicum encouraged her to find the edge of her comfort zone: an area of teaching that she didn't feel particularly confident about, document and describe her practice, paying particular attention to areas of discomfort. This way, she could enlist her peers in the practicum, as well as the practicum instructor, as her think-tank and resource people.

Celia's reflection allowed her to describe the difficulty her students had experienced in writing beyond one sentence. In her letter to her practicum instructor, she wrote,

> I must say it feels like I am hitting a block. It seems that everyone needs my help to write more. Or they would say that they don't know how to spell the words. Or they don't know what else to say. Or simply they are done—their stories are finished. Even though I can see that they can write more when I confer with them, yet I have 24 children. It is impossible for me to confer with everyone during every writing workshop. I am not sure if I expect too much too quickly? Perhaps I am not doing something right in my writing workshop?

Celia's reflection did begin by describing her observations of her students' reluctance and resistance in reaching the three-sentence goal. But she quickly took a stance as an implementer and wondered if she had set the bar too high, or if something went awry in her implementation of the writing workshop. When I, her practicum instructor, went to visit her, Celia conducted a writing workshop. After the workshop, we engaged in a conversation while the children went to lunch.

"Okay," taking in a big breath of air, Celia began the conversation half-jokingly, "Let's start with the positives."

"Sounds like you want me to give you a teaching evaluation," I chuckled, trying to envision whose voice she was imitating, her student-teaching supervisor's or her principal's?

"My principal always says that after she observes me." Celia stopped my guess work. She continued, this time seriously, "You see, most of my students are still writing just one sentence. I know that's the negative, but I want help fixing that."

The ensuing conversation with the practicum instructor allowed Celia to reexamine each steps she took in the lesson, as well as other contextual issues about her school's mandates as well as her students' histories as readers and writers, and home literacy environments. By the end of the conversation, Celia not only gained a new way of examining her teaching, but also several ideas and strategies she could use to "fix the negative." In a letter to the practicum instructor to self-reflect on her learning, Celia wrote:

> The most important thing I learned today was that I need to allow more time to sustain my students' writing efforts. Why haven't I thought about this before? It was eye-opening when you showed me how I used the workshop time: my strategy lesson took 21 minutes! I only gave my students 10 minutes to write. Only one child had a chance to share her writing before the class had to go to lunch. I am very excited that now I can try the ideas that we brainstormed to make my strategy lesson shorter so that students could have more time to write. I am also going to use the strategies that we came up with to help my students build up their writing stamina. I feel that I learned a valuable lesson today and I have a clear idea where I am going from here.

A conversation with a mentor and two letters of self-reflection later, Celia felt confident knowing that she now has several ideas to try out to help strengthen her students' writing muscles. The story serves to remind us, especially early career teachers, that reflective thinking, combined with conversations with mentors, can help us solve problems that arise as we implement new programs or use new curriculum materials. Such reflection should be regarded as cyclical in nature in order to sustain our growth overtime and prevent the hardening of ideologies.

Unlike Celia, Andrea isn't a participant in the practicum course. There is no external requirement or incentive for her to reflect on her work in writing. But two factors in her work sustain her reflective frame of mind. One of them is that she found a mentor in a fellow teacher, Mrs. Katz.

When she began teaching at her school two years ago, Mrs. Katz, a middle-aged veteran teacher whose classroom was across the hall, took her under the wing. The best part was that Andrea and Mrs. Katz took the same subway line home at the end of the day. So almost every day after work, Andrea had a thirty-five-minute subway conversation with Mrs. Katz. She found the conversation to be relaxing and fulfilling, partly because she craved adult conversations after talking to third graders all day, and partly because Mrs. Katz listened well and understood the challenges a new teacher faces.

Of many things Andrea learned from Mrs. Katz, her favorite was keeping a writing notebook to model writing. This went along with her love of writing. At first, she wrote entries to share with her students. When Mrs. Katz went on maternity leave, she began to write about her teaching. She found, in the absence of conversations with Mrs. Katz, she was now carrying on a conversation with herself in the writing notebook.

In one of her notebook entries, Andrea described the challenge in writing the Native American creation stories as follows: "Most of [my students] are fluent writers and they are not afraid to take risks in writing. So I was a bit baffled by the difficult time they are having with the creation stories…It is clear to me that I need to find a different approach to get them going." It turned out that Andrea didn't have to go far to find a different approach. One day, her trip home came to a halt as she passed the classroom of a fourth grade teacher in her school. She first noticed that her colleague was standing on a chair. Then she saw what she was looking for: a chart paper her colleague was putting on the wall. On the chart paper, in children's handwriting, she saw this:

Non-Fiction Genre Study

Book Title:
Author's Purposes:
Unique Features:

On the subway home that day, Andrea wrote in her notebook:

> Elaine's (fourth grade teacher) students were given opportunities to look through many non-fiction books and discover the genre features. I love the fact that students work in groups to come up with what they noticed from their reading. This is what I need to do to help my students study the genre of creation stories. It is the study of the genre that is missing, which explains why my kids were having a hard time writing.

Armed with the new insight, as well as a trip to her local library to borrow an armload of Native American creation stories, Andrea launched her students on a journey that began with discovery learning of the genre and ended with a publication party celebrating children's creation stories. In reflection, she wrote,

> I always knew that students learn best when they are immersed in inquiry and wonder. It shouldn't be different with the study of [Native American] creation stories. They should be given a lot of time to read many creation stories to study characteristics of this genre. Then they should be allowed

a lot of time to wonder and ask questions on their own as they create their own creation stories...This is a major risk-taking for me because I always feel the need to be in control. I am the driver in my classroom and I have to steer each student in the right path every step of the way. But too much control is not conducive to students' learning. In this case, my students taught me a lesson. As I saw my students' enthusiasm when they wrote about how popcorns or pizzas were created, I couldn't help but wonder all I did was introducing them to many creation stories and guided their discovery learning. This reflection I will take to heart in the future of my teaching.

Like Celia, Andrea used reflection to strengthen her professional expertise. Whereas Celia, being a new teacher, tended to focus on honing her teaching skills and looked to her mentors to help her walk sure-footed in how she implemented her practice, Andrea had more years of teaching experience to tap into and was able to pinpoint that she needed a different approach. The difference is in how they reflected on the experience. Celia felt that she had learned a valuable lesson from her professor, whereas Andrea credited her students for teaching her the lesson. More importantly, Andrea looked inward to examine the underlying reason—that she felt uncomfortable losing control and letting her students explore and study the genre before writing in that genre.

Strategies and Tools

We can recognize Celia's and Andrea's stories as they illustrate what a teacher thinks and does to solve problems in any given day in the classroom. You may have learned many ways to reflect on teaching and allow your reflection to guide your action. However, there is no magical way of reflection that would guarantee a spectacular solution to a problem. But it has been a century old wisdom that reflection does help us grow and become stronger. Confucius said, "By three methods we may learn wisdom: first, by reflection which is noblest; second, by imitation, which is easiest; and third, by experience, which is the bitterest."

What can we do to foster the reflective frame of mind? Reflection involves thinking about one's own conceptual and procedural knowledge, including an evaluation of the effectiveness of one's action in teaching by looking at students' responses and reactions to one's teaching. The goal of reflection is to develop reflective teachers who can monitor their teaching and their students' learning and improve the quality of teaching and learning over time. Any reflective thinking usually involves two parts: *articulation* and *reflection*.

Part One: Articulation

We don't automatically think about examining and changing the way we teach or act unless we are fully conscious of what we do. The act of articulation allows us to be aware of an area of concern. We may or may not always pinpoint exactly what the problems are. Sometimes, we just have a doubt or are perplexed by certain student behavior. Articulating the doubt or perplexity can help us put a name to it, thereby allowing us to use this state of doubt as stimulus for inquiry or investigation. This articulation can take place while speaking to a colleague or mentor, or while writing in a journal. The process of articulation may include identifying problems, describing a behavior or conflict, noting patterns or inconsistencies, acknowledging frustration or success, problematizing a situation, describing plans and intentions, explaining reasons underlying decisions and actions, and so on.

Part Two: Reflection

Reflection on the basis of articulation allows us to stretch our thinking and to make our problem-solving muscle strong and flexible—an important attribute for professional development. Reflection without first articulating often results in retelling or detached speculations. Reflection based on the articulation can be more focused. Like articulation, reflection can take place in a collegial conversation or in a journal. Reflection can include exploring possible solutions to problems, looking for underlying or contextual causes for certain behavior or conflict, explaining certain patterns or inconsistencies, exploring causes for frustration or success, problematizing what was taken for granted, or raising questions about plans, intentions, decisions, and actions.

The biggest challenge for effective reflection is lack of time. After all, how many of us reflect because we are required to do so by a graduate course? How many of us are lucky enough to have a thirty-five-minute end-of-the-day conversation with a mentor teacher on a daily basis? How many of us can fit journal writing in our busy and demanding world of teachers (and parents, spouses, children to aging parents, pet owners)? Sure, we face the same challenge with finding time to do many other important things in life, like exercise. We can admire someone else's arm with chiseled muscles and say, "Oh, if only I have the time, I'd pump iron in a gym." However, we know they don't get those muscles because they exercise once in a while. The same is true with reflection—occasional reflection doesn't help you build your problem-solving muscles. There is no easy way around it. Here are some strategies for reflection that you may find easy to fit into your busy lives.

Reflective Checklist

There is a plethora of checklists from various sources, from professional books to your school. Some new teachers receive checklists in the welcome packets as they attend the school staff meeting for the first time. If you are a new teacher, you may find that using a checklist can help you get organized. But how can you use it as a tool for reflection? When you use it to examine your realities and articulate your concerns, it can be a tool for reflection. This means that you need to use a checklist that works for you. If not, you need to revise it or create one that allows you to articulate your concerns, thus leading you to reflection. For example, a checklist of questions to help you reflect on your use of classroom environment may include questions such as these:

- Does my classroom arrangement encourage students to become independent problem-solvers?
- Do I provide a variety of resources in my classroom to meet the needs of all my students?
- Does the arrangement of my class library encourage children to pay attention to genres, authors, subject areas, as well as reading levels?
- Do I encourage and enable students to access and use a wide variety of resources?
- Do the print materials on the walls represent processed as well as commercial information?
- Does my classroom environment encourage students to develop ownership of their learning?

Questions on this checklist force you to provide a "yes" or "no" or "maybe" for an answer without much reflection. They already assume that you will comply without thoroughly examine your classroom. To make the checklist work for you, and perhaps more importantly, to think more reflectively, you need to run them through a two-column table. In the first column, you can list the questions from the checklist. The second column is where you adopt a more reflective stance and rephrase the questions to bring them closer to your own classroom realities. For example, the question "Does my classroom arrangement encourage students to become independent problem solvers?" can be rephrased as "How is my classroom arranged? What role does the room arrangement play on the students' problem-solving process? To what degree is my classroom set up to encourage students to become independent problem solvers?" For a list of complete rephrased questions on classroom organization, please refer to *Rephrased Checklist* in the Tools for Practice section.

A Blank Space

In my early years as a teacher, I designed a lesson plan format. I drew a line four-fifths down a page. On the top portion, I wrote my lesson plan. After each lesson, I'd usually go back to the lesson plan, and fill out the bottom portion with my reflection. I would jot down my thoughts on the lesson, including my flounders and good moves, as well as students' responses. For me, that blank space at the bottom of the lesson plan is a standing invitation for me to be thoughtful of my own teaching.

I used this format until I found Post-its, which afforded me opportunities to be thoughtful beyond my own lesson plans. So consider using

Post-its as an alternative blank space that you can carry with you and literally attach your reflection on anything that comes to your attention: your own lesson plans, books you plan to read aloud or have read aloud, and children's conference record. The possibilities are endless. Having a stack of Post-its at our fingertips makes it easy to jot down our reflective thinking while we are on the go. This is the busy teacher's way of leaving footprints in the sand, or in Donald Schon's term (1987), "reflection-in-action."

A Double-Entry Journal

When my daughter started first grade, her morning routines were often chaotic. Several times before she got on the bus, she would suddenly turn to me and ruefully exclaim, "I forgot my lunch ticket!" or "I forgot to wear my sneakers for gym!" To help her become organized, I put up a checklist on the refrigerator that included lunch ticket, water, snack, gym shoes (Tuesday and Thursday). My daughter dutifully checked the list every morning, until early October, when I forgot to print out a fresh copy of the checklist. But my first grader proudly announced that she didn't need one any more. To prove this, she yelped, "Lunch ticket, check! Water, check! Snack, check! Gym shoes, X because it is Monday!"

A double-entry journal, in a way, serves the same purpose as my daughter's checklist. Once you have developed a reflective frame of mind, you do not necessarily have to dutifully write in it everyday. Writing helps you make your conceptual and procedural thinking visible. You should, ideally, always carry a double-entry journal in your head. You need to find a notebook that works for you. Some teachers like a small spiral notebook so it can be easily carried around and used on a lap when you don't have access to a desk. Others use the marble top notebooks, which can double up as a writer's notebook, like Andrea's notebook.

If the page of the notebook is wide enough, you can divide the page in half. On top of the left column, write Articulation. Save the right column for Reflection. If you don't have time to complete both columns in one sitting, you can just fill out the Articulation column and leave the Reflection column for later. You may find that whether or not your reflection can help you solve problems or contradiction is not as important as the fact that you noticed the problems and articulated them. Please refer to the Tool's section at the end of the chapter for an example.

Some early career teachers we work with swear by the double-entry journal. They find it hard to be deeply reflective while juggling teaching, managing, and supervising. They may agree with the Taoist saying that "No one can see their reflection in running water. It is only in still water that we can see." This seems to be in agreement with Donald Schon's notion of "reflection-on-action." For some teachers, having reflective Post-its everywhere is unappealing and chaotic. If you are one of them, give the double-entry journal a try.

Revisit Reflection

Reflection of your teaching involves, at least in part, a revisit of your teaching. What about a revisit of your reflection? A revisit of your reflection is simply just that—carve out a time in a week, or a month, in which you can reread your Blank Space thoughts on the lesson plan, your reflective Post-its, or your double-entry journal. You may consider writing marginal notes as you reread, or create a third column in your double-entry journal to jot down your observations and comments.

The question at the heart of this revisit is: What does this say about me as a learner/teacher? The revisit allows us to stand on the shoulders of our own thinking. During the revisit, we look for connections between theory and practice, between best practice and our practice; we also look for patterns and themes emerging in our reflection; we contemplate our imperfections as well as taking stock, allowing us to feel rewarded and encouraged by victories big and small; we develop lines of inquiry for exploration. Above all, we get in touch with who we are as teachers and learners. The learning needs for you and your students are always paramount. And remember not to shoehorn your reflection into any narrow parameters.

Three Levels of Reflective Stance

Once you begin to benefit from a reflective stance, it's time to consider how to improve the quality of your reflection. For the purpose of helping new teachers, especially early career teachers, to go beyond immediate concern about teaching practice, we describe three levels of reflection based on three roles that teachers must play, often simultaneously (figure 3.1): implementer reflection, observer reflection, and transformer reflection.

Implementer reflection. Reflection at this level is largely concerned with the processes and outcome of implementation of curriculum materials, units of study, lessons, strategies, group work, and so on. In this type of reflection, the teacher/implementer usually looks back and evaluates her action retrospectively. The questions associated with this level of reflection are "How did it go?" "Did I do it right?" "What went well?" "What didn't go so well?" Often

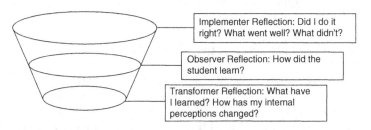

Figure 3.1   Three levels of reflection.

at this level, reflection aims at improving one's ability to carry out teaching tasks, without explicit concern about the qualities of students' learning.

Observer reflection. The term suggests that reflection at this level focuses on keeping a watchful eye on student learning as an outcome of one's own teaching. Reflection of this type can take place during and after teaching. Often the teacher/observer documents and describes the students' learning, especially those who seem to struggle, in order to come up with appropriate responses to help them. The questions associated with this level of reflection are: "How did my students learn?" "Why did Diego refuse to do any revision?" "Why did this literature circle take so long to finish a book?" Questions like these lead the teacher/observer to explore ways to respond to students' needs and to solve problems, which is at the heart of this level of reflection.

Transformer reflection. The focus at this level of reflection is on considering and reconsidering various perspectives, on making connections between these perspectives and one's own, between theory and practice, and between past and future practices, leading to fundamental changes of one's own perception as well as practice. The teacher/transformer looks for connections, questions assumptions, looks for changes to take place, asking, "How have I learned?" "What new insights have I gained?" "How did the new information lead to a change in my internal perception?" This level of reflection often leads to professional and personal growth and rejuvenation.

We believe that these three levels of reflection are all important but perhaps they are developmental, as a teacher with little teaching experience to draw on would tend to focus more on how she implemented learning activities; whereas a teacher who has mastered routines and procedures and has implemented various curriculum and activities would be more likely to observe how the students are learning, and what she needs to change in order to sustain professional growth.

In air travel, we are reminded that in an emergency situation that requires the use of oxygen, we should put oxygen masks on ourselves before helping others. Teaching without self-reflection is like trying to help others while you are running out of breath. Reflection is a self-renewal process. It allows us to put the oxygen masks on ourselves so that we can help others more effectively.

Tool for Practice: *Reflective Checklist*

Use this tool to run any checklist under the reflective lens. List the checklist questions in the left column. Then in the right column, rephrase the questions so that you are more reflective beyond simply answering "yes" or "no." For example, a checklist for classroom environment can become be framed as shown in table 3.1.

Table 3.1   Checklist for classroom environment

Checklist questions	Rephrased questions
Does my classroom arrangement encourage students to become independent problem solvers?	How is my classroom arranged? What role does the room arrangement play on the students' problem-solving process? To what degree is my classroom set up to encourage students to become independent problem solvers?
Do I provide a variety of resources in my classroom to meet the needs of all my students?	What resources have I provided in my classroom? What are my students' needs? To what degree do my classroom resources meet the needs of all my students?
Does the arrangement of my class library encourage children to pay attention to genres, authors, subject areas, as well as reading levels?	How do I arrange my class library? How can I arrange my class library so that children can pay attention to genres, authors, subject areas, as well as reading levels? How can I get my students involved in arranging the library so that they will pay attention to these areas?
Do I encourage and enable students to access and use a wide variety of resources?	To what degree do my students access and use a variety of resources? How do I encourage them to have a wider access to a variety of resources?
Do the print materials on the walls represent processed as well as commercial information?	What print materials are on my classroom walls? What do I want my students to get out of the print materials? How can I get students involved in creating and contributing to some print materials on the walls?
Does my classroom environment encourage students to develop ownership of their learning?	To what degree does my classroom encourage student to develop ownership of their learning? How can I involve students in creating a learning environment that fosters their sense of ownership for their learning?

Table 3.2   Double-entry journal

Articulation	Reflection
Identifying problems	Exploring possible solutions to problems
Describing a behavior or conflict	looking for underlying or contextual causes for certain behavior or conflict
Noting patterns or inconsistencies	
Acknowledging frustration or success	Explaining certain patterns or inconsistencies
Describing plans and intentions	Exploring causes for frustration or success
Explaining reasons underlying decisions and actions	Problematizing what was taken for granted
	Raising questions about plans, intentions, decisions, and actions

Tool for Practice: A Double-Entry Journal

Reflective process or thinking usually begins with articulation—identify problems, describe noticings, or voice your concerns. Using this tool allows you to build your reflective thinking on articulated realities (table 3.2).

Table 3.3   Revisit reflection

Articulation	Reflection	Revisit reflection
Identifying problems Describing a behavior or conflict Noting patterns or inconsistencies Acknowledging frustration or success Describing plans and intentions Explaining reasons underlying decisions and actions	Exploring possible solutions to problems looking for underlying or contextual causes for certain behavior or conflict Explaining certain patterns or inconsistencies Exploring causes for frustration or success Problematizing what was taken for granted Raising questions about plans, intentions, decisions, and actions	Looking for connections between theory and practice Looking for connections between best practice and our practice Looking for patterns and themes emerging in our reflection Contemplating our imperfections Taking stock Developing lines of inquiry for exploration Getting in touch with who we are as teachers and learners

Table 3.4   Three levels of reflective stance

Levels	Implementer reflection	Observer reflection	Transformer reflection
Focus questions	Did I do it right? What went well? What didn't?	How did the student learn?	What have I learned? How has my internal perceptions changed?
Areas of concern	Technicalities, routines, and procedures of instruction	Student learning processes and products	My own learning as a teacher as well as a learner

Tool for Practice: *Revisit Reflection*

By adding a column to the double-entry journal, we allow ourselves to stand on the shoulders of our own reflection thus bringing our reflective stance to a new level, as shown in table 3.3. Or you can simply make marginal notes as you reread your double-entry journal.

Tool for Practice: *Three Levels of Reflective Stance*

This tool allows you to raise your awareness of the three levels of reflection. It can also help you examine the quality of your reflection. Reread your reflection, whether you keep a teaching journal or double-entry journal, and check to see if your reflection tends to be of one kind, or if you have a variety of reflections. If you are a new teacher and your reflection tends to lean more toward Implementer Reflection, then you might want to pay more attention to how your students are learning. If you have very little Transformer Reflection, then you may want to look inward and pay more attention on your own learning and growing.

Suggested Resources

Bransford, J.D., Brown, A.L., & Cocking, R.R. (Eds). (1999). *How people learn: Brain, mind, experience and school.* Washington, DC: National Academy Press.

Brunner, D.D. (1994). *Inquiry and reflection: Framing narrative practice in education.* Albany, NY: State University of New York Press.

Cruikshank, D.R. (1985). Uses and benefits of reflective teaching. *Phi Delta Kappan,* 66(10), 704–706.

Schon, D. (1983). *The reflective practitioner: How professionals think in action.* New York: Basic Books.

———. (1987). *Educating the reflective practitioner.* San Francisco: Jossey-Bass.

Sparks-Langer, G.M., Simmons, J.M., Pasch, M., Colton, A., & Starko, A. (1990). Reflective pedagogical thinking: How can we promote it and measure it? *Journal of Teacher Education, 41,* 23–32.

Wedman, J., & Martin, M. (1986). Exploring the development of reflective thinking though journal writing. *Reading Improvement, 23*(1), 68–71.

References

Clark, A., (1995). Professional development in practicum settings: Reflective practice under scrutiny. *Teaching and Teacher Education, 11*(3), 243–261.

Collier, S.T. (1999). Characteristics of effective teachers. *Journal of Teacher Education, 50*(3), 173–181.

Darling-Hammond, L., & Snyder, J. (2000). Authentic assessment of teaching in context. *Teaching and Teacher Education, 16*(5), 523–545.

Dewey, J. (1933). *How we think: A restatement of the relation of reflective thinking to the educative process.* New York: D.C. Heath and Company.

Hatton, N., & Smith, D. (1995). Reflection in teacher education: Towards definition and implementation. *Teacher and Teacher Education, 11*(1), 33–49.

Korthagen, F.A.J. (1999). Linking reflection and technical competence: The logbook as an instrument in teacher education. *European Journal of Teacher Education, 22*(2/3), 191–207.

Loughran, J.J. (2002). Effective reflective practice: In search of meaning in learning about teaching. *Journal of Teacher Education, 53*(1), 33–43.

Rearick, M.L., & Feldman, A. (1998). Orientations, purposes and reflection: A framework for understanding action research. *Teaching and Teacher Education, 15*(4), 333–349.

Reiman, A.J. (1999). The evolution of the social role-taking and guided reflection framework in teacher education: Recent theory and quantitative synthesis of research. *Teaching and Teacher Education, 15*(6), 597–612.

Risko, V., Vukelich, C., & Roskos, K. (2002). Preparing teachers for reflective practice: Intentions, contradictions, and possibilities. *Language Arts, 80*(2), 134–144.

Schon, D. (1983). *The reflective practitioner: How professionals think in action.* New York: Basic Books.

———. (1987). *Educating the reflective practitioner.* San Francisco: Jossey-Bass.

Stanley, C. (1998). A framework for teacher reflectivity. *TESOL Quarterly, 32*(3), 584–591.

Valli, L. (1997). Listening to other voices: A description of teacher reflection in the United States. *Peabody Journal of Education, 72*(1), 67–88.

Van Manen, M. (1977). Linking ways of knowing with ways of being practical. *Curriculum Inquiry, 6,* 205–228.

Ward, J.R., & McCotter, S.S. (2004). Reflection as a visible outcome for preservice teachers. *Teaching and Teacher Education, 20*(3), 243–257.

Zeichner, K.M. (1996). Teachers as reflective practitioners and the democratization of school reform. In K.M Zeichner, S. Melnick, & M.L. Gomez (eds), *Currents of reform in preservice teacher education* (pp. 199–214). Teachers College Press, New York.

4

Observing the Lesson

When contemplating the upcoming visit, there are many things on my mind. Although I'm not exactly nervous, I'm not entirely comfortable either. It can always be a little nerve wracking to have an outside visitor come in, regardless of how comfortable I may be with my current teaching practices.

—(Diane, a third-year teacher, reflecting about the impending visit from her graduate school professor)

Diane articulates a unique discomfort many teachers experience when they anticipate a visit by an outsider to their classroom. While teachers spend hours each day in front of their students, it is rare to have another adult in the room, observing and providing feedback about the lesson.

Challenge and Opportunity

The "Observation" is one of the benchmarks of a teacher's experience. From your time as a pre-service teacher through your professional career a teaching observation is often anxiety-ridden, an experience to be endured rather than enjoyed. For many teachers an observation has negative connotations, as teachers frequently strive for perfection and focus on their performance. The challenge is to see how the observation experience can be reclaimed as an opportunity for growth rather than an unwelcome experience. And in reality many observations, especially by supervisors, are evaluative and serve as a summative judgment on teaching. Yet what if an observation can be used as an opportunity for growth? What kinds of professional learning can occur when teachers can become more open to having supportive observations? In reclaiming the observation process, we offer teachers an opportunity to take some control of the observation by framing questions—rather than have the observer solely determine its use.

Similarly, we also see a challenge in asking teachers to observe each other to create learning opportunities to foster growth instead of competition. While it can be uncomfortable to watch yourself on video, the experience offers a unique opportunity to view your teaching from a different perspective. In busy teaching lives, there is little time to watch other teachers. We may not know what is going on down the hall, not to mention in other schools. This can create a feeling of isolation and increase your concern about how you are doing.

We define the observation process as having your own teaching watched, viewing your own teaching through video, and learning how to view another teacher's work. Defined in this way, observation can provide invaluable opportunities for professional growth. In our practicum course, we created a context to enable new teachers to take ownership of the observation process, rather than to fear it. We facilitated this in several ways. First, within the seminar of the practicum we talked openly with the teachers about our goals to explore and stretch their teaching comfort zones. Giving teachers an opportunity to air their fears was the first step.

The course activities scaffold exploring the observation process for the teachers. As the course instructors, we visit each teacher at his or her school. We see our roles as coach and mentor, emphasizing that this is an opportunity to learn. Prior to the visit, the teachers are asked to write a letter of invitation to the instructor, identifying an area in their literacy education that they are eager to explore, and describe the lesson/activity/task that would transpire during the visit. We encourage them to ask questions and describe their uncertainties or specify how they want the instructor to help them. We meet with the teachers after the visit to share our reactions. After the visit, the teacher sends a letter to the instructor, summarizing the visit and reflecting on any insights or changes resulting from the visit.

We also facilitate learning to watch and coach peers. Teachers capture their learning by making a videotape of peers in action in the classroom. They then exchange the videos, along with their self-assessment and critique of the video with a peer. After viewing the tape, teachers write a letter of response to their peer. Undoubtedly observations have and will continue to play a role in your teaching life. In this chapter we invite you, by sharing the stories of teachers like yourself, to examine your experiences and insecurities about being observed and consider ways in which watching the lesson can become a catalyst for your professional growth.

Stories from the Field

Anticipating a Visit from a College Professor

Nicole loved to share stories about her third grade students. As a second-year teacher, she felt she had a strong grasp of her reading instruction and was confident in her teaching. Yet she was concerned about the Writing Workshop. While she understood the components of the workshop structure, she felt overwhelmed in her attempts to manage her students' varying needs. Yet she was reluctant to ask for help in her own school. Like many teachers, she believed she needed to solve her problems herself. If she asked for help from her colleagues or from a supervisor, they might think she was incapable or that she did not know what she was doing. The thought shook her confidence.

As a participant in our practicum, Nicole had the opportunity to address her concerns as she contemplated the site visit. In her letter of invitation she wrote:

> The independent writing time is quite difficult for quite a few of my students. They struggle with following directions and getting started. I've been wondering about how to address these different needs in the Teachers College workshop format. Last year, I do not feel that I served the strugglers as well as I would have liked. It doesn't seem fair or possible to spend all of my time with these students without attending to those who are attempting to use those strategies. It is during this conference time that I need support. How can I manage the classroom space better when conferring? What can I do to better support those strugglers in this space? How can I differentiate the workshop?

Writing this letter gave Nicole an opportunity to articulate her questions. She was able to confront and articulate a part of teaching that made her anxious and was able to ask for help in a safe way. Still, she felt trepidation about the upcoming visit.

Watching Myself Teach

"No way!" Sandra exclaimed when we talked about the peer videotaping assignment. As a third-year teacher in a private school, Sandra spoke confidently about her teaching strategies; so her strong negative reaction to the videotape project was surprising. Yet as we talked, it became clear that it was because Sandra believed she was a successful teacher that she feared anything that might change what was already working. "If it isn't broken, don't fix it" seemed to be her motto. Videotaping herself, for her own evaluation and that of a peer, threatened her hard-earned sense of self.

Despite her reservations, Sandra videotaped herself leading a lesson on retellings. Writing a letter to her peer observer, she comments on her own performance and asks for advice:

> I felt like I was rushing through the lesson. I feel that there were so many things that were different about the lesson it's hard for me to pinpoint things to look for. I know I went quickly—perhaps you can give me some pointers that will help me with my pacing. Also I am still struggling with my questioning and language usage when I talk to my students. I previously taught in the upper grades and still don't feel comfortable with giving directions to my class.

How do you feel about videotaping your own teaching? Do Sandra's reflections about her performance resonate with you?

Preparing to Coach a Peer

Kara teaches in a third grade charter school. She was concerned about being "smart" enough to give another peer any advice. As a first-year teacher she did not believe she could offer any comments to another teacher. Without an on-site coach or extensive professional development at her school, she felt very much on her own as she developed her literacy curriculum. However, she was extremely curious to see another teacher's third grade classroom. With so much to attend to in her own classroom, with the needs of her own children, she rarely had the chance to see what other teachers did. "I feel like I live in a teacher bubble," she wistfully remarked. "I'm not sure I know what to say to my partner."

How frequently do you get an opportunity to move out of your "teacher bubble"? What do you think you could learn from watching a colleague's teaching? We believe that observing others as they teach offers a wealth of opportunities to develop insight into practice. What might be challenging about sharing your ideas about what you observe?

Research and Reflection

Every education program course of study has clinical courses. The prevalent model of teacher education requires that pre-service teachers spend time in classrooms in order to observe, learn, and have opportunities to teach, with supervision. We all come to teacher preparation programs with expectations and understandings about teaching, drawn from our own experiences (Lortie, 1975; McDiarmid, 1999); so it is important that experiences in the student teaching courses help pre-service teachers recognize and address their preconceptions about teaching. Field observations, then, are best when they are guided by teacher educators in order to help pre-service teachers notice specific aspects of classroom practice. It is important that connections are made to research about best practices and that the observations are structured to make sure that pre-service teachers are directed toward viewing exemplary practice. Sabers et al. (1991) showed videotapes of reading lessons to novice teachers and experienced teachers and asked them to think aloud about they saw, with further questions from researchers to probe their thinking. Expert teachers were able to discuss and interpret the videotapes differently than the novice teachers. They found that novice teachers focused on isolated events and surface structures of the interactions and that the experienced teachers were able to see the bigger picture and larger goals. Roehrig et al. (2008) investigated the impact of a pre-service literacy course that included explicit instruction and experiences in focusing students' attention to exemplary teaching practices that promoted children's literacy engagement and achievement. They found that by focusing pre-service teachers' attention on what to look for they were able to identify and then emulate effective literacy practices.

In addition to learning how to observe teachers during student teaching experiences, pre-service teachers are observed as part of their clinical experience. Typically this model follows three steps: a pre-conference between the student teacher and the supervisor, an observation in the classroom, and a post-conference. This model frequently continues into the school arena where observations are done by administrators to evaluate teachers' performance. Vogt and Shearer (2007) state from the 1970s onward, the model for teacher evaluation was one where administrators were presumed to be experts. They would briefly observe a lesson, tell teachers what worked, and what didn't work. The administrators would do most of the talking. There would be a written evaluation placed in the file. Yet with all the demands on their time, administrators may not necessarily be well versed in current literacy instruction. Their feedback to teachers then won't provide teachers with needed, specific literacy leadership (Henk et al., 2000). With the unequal power relationship between teacher and administrator, this model doesn't promote collegiality or professional development.

Literacy coaching, which includes lesson observation and feedback, has become an important vehicle in supporting teachers' professional development. With an ever-evolving role in schools, their goal, according to Walpole and Blamey (2008), is to "build teacher knowledge so that children's literacy achievement will be increased" (p. 222). As such, their observations, as Casey (2006), Toll (2005), and Allen (2006) explain, is one of collaboration and trust to build upon strengths.

Finally, observation tools, such as the Reading Lesson Observation Framework (Henk et al., 2000), can be used by schools and districts to describe and then facilitate the observation of reading instruction. This instrument describes, in great detail, components of a reading lesson, and provides the observer with a focused checklist with a rubric. Such a framework will, the creators argue, "increase cooperation and communication among literacy educators within a school district and bring them to some much needed common ground" (p. 367).

Videotaping

Videotaping, as part of teacher education programs, is a continually evolving process. It originated in the late 1960s to provide more effective feedback to student teachers (Frieberg & Waxman, 1988). Extensive research supports the use of videotaping teachers as they progress from student teacher to novice teacher to experienced teacher (Acheson & Gall, 1987; Armstrong, 1999; Showers, 1985; Song & Catapano, 2008). Armstrong (1999) examined how videotaping student teachers during their field placements led to more focused and perceptive comments on their performance, especially in the area of classroom management. With the use of multiple videotaping opportunities, student teachers were, over time, able to shift their attention from their own personal idiosyncrasies to issues of pedagogy and children's achievement. Song and Catapano (2008) surveyed in-service urban teachers about the impact of videotaping on their practice and also used external videotape reviewers to assess the teaching practice. They found significant growth in the teaching practices. Finally, Dicks (2005) shared how using videotapes of both students and his own teaching served as a powerful catalyst to students and to his own professional learning. He stated, "I have witnessed a dramatic increase in student learning through the use of digital video feedback. By becoming catalysts in developing their own strategies, my students have grown more confident and competent in their communication skills" (p. 79).

Others have raised cautions about the proliferation of videotaping protocols in teacher education programs. Greenwalt (2008) interviewed student teachers who completed a student teaching video assignment. Some stated that the presence of the camera altered their teaching, causing them to be overly self-conscious about their bodies and their voices, rather than on their teaching. They saw the camera as an intrusion on their teaching. They were concerned that the camera wouldn't capture the reality of the classroom and its context. Smith (1996) reported that in-service teachers, required to use videotaping in graduate courses, experienced difficulties with the technical aspects of the assignment as well as expressed concern about how they would be perceived on the videotapes. These participants rated videotaping as the least useful and least wanted teacher observation tool. To counter this, Smith argues that teachers need to have greater control over the taping experience, discussion, and coaching through the experience and their feelings about watching themselves, and opportunities to share the experience with others.

Videotaping, used purposefully, thoughtfully, and most importantly, collaboratively, offers a unique opportunity for careful study of practice. As a literacy coach, Casey (2006) argues that without videotaping we are not able to fully get the perspective needed to improve practice. To support and encourage teachers, she videotapes herself before asking teachers to videotape themselves. She also models using the videotape as a way to make instructional changes. She videotapes herself teaching a lesson, then, along with teachers, identifies the strengths and areas of need in the lesson. Together they create an action plan, with specific changes to try. She then tapes a second lesson, incorporating the changes. This practice models a way to use the videotape to analyze instruction, create an action plan, and then carry out the plan.

Bursting the Teaching Isolation Bubble

If you think about your teaching day, how often are you able to seek the support and guidance of your peers and school leadership? Teachers are frequently isolated in their classrooms, with little opportunity to visit or interact with their peers. This can lead to a belief that teaching is private. Fullan and Hargreaves (1996) describe isolation and privatism in schools as "deep seated. Architecture often supports it. The timetable reinforces it. Overload sustains it. History legitimates it" (p. 6). Isolation can make it difficult for you to grow in your practice (Gallimore & Tharp, 1990; Swafford, 2003).

In the past professional development was frequently regulated to a workshop here and there, usually to present a new program or technique. Joyce and Showers (2002) argue that effective professional development or learning is not simply a workshop or a book or a website; it is embedded within teachers' day-to-day work. Additionally, older models of professional development, such as weekly staff meetings or large, one-day workshops, did not necessarily promote deeper content understanding, collaboration between teachers, or lasting professional growth. The increase in implementation of coaching since the 1990s is a professional development movement that is showing promise in changing the isolation of teachers. Darling-Hammond and McLaughlin (1995) argue that the essential features of effective professional development include collaboration and sharing of knowledge among teachers; modeling, coaching, and collective problem solving; and engagement of teachers in the authentic, concrete tasks of teaching. The U.S. Department of Education (2001) found that high-quality professional development effectively prepares teachers for challenges when it is of sufficient length, frequency, and intensity; involves supporting teachers to help move their students to state standards of achievement; gives teachers a primary role in planning; and provides time to practice new strategies.

New teachers especially benefit from professional development opportunities that involve coaching. There is no "one size fits all" for professional development as beginning teachers are not all the same (Valencia et al., 2006). Effective professional development is best found at the school level, where there are both formal and informal interactions between teachers and coaches. Valencia et al. (2006) report that the beginning teachers in their study, when asked what they would most want, would choose "someone on-site to help." They continue, "These novice teachers wanted to open their classrooms to expert teachers who would watch them teach, provide feedback, and model lessons" (p. 117).

Effective professional development, then, must be grounded in the actual teaching work that occurs day in and day out. As Dearman and Alber (2005) state, "teacher learning is most effective in the context of teaching" (p. 637). Participating in intravisitations (visits to other teachers in your schools) and intervisitations (visits to teachers in other schools) are ways to observe other teachers in order to improve your instruction (Casey, 2006).

Swafford (2003) studied a group of early career teachers who were involved in a variety of peer coaching activities, such as group discussions, shared viewing of participants' teaching videotapes, and visits to each other's classrooms. Through interviews with the teachers, Swafford learned that the peer coaching activities supported teachers in three ways. Initially, the coaching support was procedural, a focus on the technical aspects of teaching. This could be assistance in selecting materials, help in planning a mini lesson, or classroom organizational techniques. Another kind of support was affective. Peer coaching helped alleviate some of the personal doubts teachers had about their performance, enabled teachers to ask questions about areas for improvement, and facilitated teachers in the willingness to take risks. Finally, the peer coaching provided a conversational space for reflection about literacy teaching. In these conversations teachers could move beyond the details of the lessons to consider their strengths and needs as literacy professionals, and to anticipate professional growth.

As the research shows, there are many ways to observe lessons. You can invite a colleague or coach to observe you as you work through an aspect of your practice. You can use videotape to get a new perspective on your own teaching and your students' responses. And you can learn by observing your colleagues and by participating as a peer coach. All of these activities will support your growth as a teacher and help you move out of your teaching bubble. As Fullan (2000) argues, it is important to open your classroom (and your mind) in order to engage in questioning and thinking, and investigating practice with peers is a powerful way to grow as teachers.

Revisiting the Field

Making Changes to Practice

I visited Nicole's classroom a few days after I received her letter. As much as possible I aim to make my visit informal and unobtrusive. I enter the classroom early, find time to informally check with Nicole, and introduce myself to the children. I ask Nicole if one of the students could show me around the classroom. This gives me an opportunity to learn about the context and helps the children and Nicole feel more at ease with my presence.

During the lesson I observe both Nicole and her students, using her questions about her struggling readers as a focus. While I take a few notes, as demonstrated in the Tool for Practice *Viewing a Colleague's Lesson*, I make sure to focus my attention on Nicole and her students.

During the session Nicole teaches a mini lesson about showing rather than telling in writing personal narratives. Many of her students wrote simple sentences such as "I felt sad" or "She was happy." She wanted them to be able to revise their sentences to move beyond the adjectives to adding details to show the reader how the character felt. Her students, seated on a rug, giggle in delight and recognition as she models writing with a childhood story about her jealousy of her big sister. After about ten minutes, she sends them off to find a spot in their own work where they can use more details, rather than a single adjective. She then begins to conference with several students, while keeping an eye on the other students. I sit in on a conference and then visit with children, checking in on how they have understood and applied the mini lesson. After about fifteen minutes, many of the children have stopped writing and started chatting; and Nicole frequently interrupts her conferring to gently remind her students to continue their writing.

Soon the music teacher comes in and Nicole and I find a corner of the school library to debrief and confer about the lesson. Just as in a writing conference with a student, it is vital to invite the student to do most of the talking. I invite Nicole to share her first thoughts before I share my ideas. She immediately speaks about the management and expresses disappointment in her students' beginning to talk before the end of writer's workshop. I take that opportunity to point out how much attention they had given their writing, from their complete engagement in her skillful and well-paced mini lesson to a solid fifteen minutes of writing. I reminded her that this was September, and that as third graders they were building their writing stamina. We then talk about ways she could foster that stamina, how she could also help struggling writers make the transition from the mini lesson at the rug to their independent work, and how she might collect data from her students' writing ahead of time so that she could more systematically target her students' needs during the conferences. The overarching structure of my conference with Nicole follows one typical of literacy coaches (such as Casey, 2006; Toll, 2005) by beginning with strengths and then shifting to areas to work on. My comments were meant as suggestions to try, not mandates, and Nicole felt excited about trying some new, practical ideas.

Following my visit to Nicole's third grade writing workshop, she was pleased to report on the progress she had made:

> After your visit and our subsequent conversation, there was much to think about. In order to do my best job as a writing teacher/coach, I need to be mostly still during the workshop time. Having Post-its while I confer with students is helpful in that it decreases the likelihood that they will forget what we had discussed as they move from the rug (with me) to their seat. I like the idea of taking home several of the students writing folders and jotting notes for them so that they have a definite task to complete or think about during some portion of the workshop period. The topic of the Post-it would vary according to the needs of the child.
>
> A couple of days after your visit, students finished revising their first personal narratives. I took their pieces home over the holiday to get a true glimpse of their abilities as writers. As suggested, I created a spreadsheet entitled "Writing Traits" and identified the traits that I noticed in each child's piece. It was reaffirming to see that the strategies introduced in class had been successfully used by the majority of the class.
>
> Of course, looking at this spreadsheet there are also areas that need to be readdressed with some of the students (i.e. using dialogue, showing emotion, etc.) or which students are ready to be pushed to the next level of a specific strategy (i.e. using quotation marks). During the next personal narrative unit, I now know which students need to be pulled into specific skill/ strategy groups. I am very excited about this, as I have not held many small group conferences.

Nicole's letter demonstrates the learning that can take place with a teacher-initiated observation. She changed several procedures during the writing workshop, such as staying in one spot and using Post-its to help the children focus on their independent work. Additionally, she experimented with creating an assessment system so that she could have a stronger, clearer grasp of where the children were with their writing. Implementing these changes helped her see the effectiveness of her teaching more clearly and enabled her to move forward in her instruction. Most importantly, the changes generated an excitement about teaching writing in new ways.

Viewing the Videotape

Despite her initial reservations about videotaping herself, Sandra embraced the project and decided to videotape herself teaching a small-group writing lesson, an area she considered weak. Here is her response following her video exchange:

> My video was very challenging for me because I decided to tape myself in an area that I am not so confident in, and feel I have trouble with. However, after viewing the video and hearing positive feedback, I actually saw things about the lesson that I liked! It was great to have another eye providing feedback for me in a nonjudgmental way. I also learned from my partner how to change my teaching and be more explicit, and I think I am a better teacher after reflecting on this lesson. I am now eager to share concerns and anxieties with my colleagues because I am not as afraid and I am much more confident; this change has enabled me to receive lots of great advice and tips. I feel I have also grown as an educator, because I feel I am more apt to take risks this semester in my classroom and try new ideas out. I now realize that the most important thing about teaching is reflection; reflection on the lesson and reflection of myself as a teacher. I learn from myself and my students everyday, and I see how taking the time to reflect each day helps me fine tune activities and meet my students needs as their teacher.

Sandra has found that watching herself on videotape and inviting a colleague to respond directly to her concerns has had a powerful effect on her teaching and her sense of herself as a teacher. By explicitly addressing her concerns she was able to get clear feedback. Receiving specific positive feedback as well, Sandra was also able to reaffirm her confidence in her abilities and in the observation process. By extending her comfort zone and watching herself carefully on video, she has moved forward in her professional development as a teacher.

Learning through Giving Feedback

After completing the video exchange assignment, Kara remarked that she had watched Laura's video more than five times because she wanted to be sure to do a good job in providing feedback. The energy and attention devoted to such close observation helped her see the nuances in the lesson and enabled her to carefully craft a response that recognized Laura's strong organization, her enthusiastic delivery, and the skillful use of visuals. She was also able to offer a few suggestions about the text and how she might engage some of the restless children. Kara realized the value of watching another teacher teach:

> It is so helpful to actually see each other teaching. I always find it helpful to pick up little ideas here and there from other teachers. It was also helpful to see Laura doing some of the same things that I do while teaching. Getting out of my teaching bubble to observe other teachers is always something I need to remember to do.

Leaving her "teaching bubble" helped Kara recognize her own strengths and learn new strategies.

Strategies and Tools

In the previous sections we've shared the experiences of teachers who have stepped out of their teaching bubbles, as discomforting as that can be, to learn to be observed, observe themselves, and observe others. These experiences are not "one shot" looks, but observations embedded in continual and collaborative conversations. As you read the next section and contemplate trying these activities, it is useful to identify a colleague, trusted mentor, or friend from another school, to be your thinking partner or critical friend. Working with someone you trust and respect will make this a unique and powerful experience.

Professional Growth through Observations

Our experiences with teachers who take ownership of the observation process convinced us of the importance of closely watching lessons. These experiences provide catalysts for professional growth by helping teachers acknowledge and take ownership of their teaching strengths and weaknesses, providing a forum for teachers to ask for and receive targeted information and support. Here we share the stories of teachers as they reclaim the observation process to further their own professional growth. We provide Tools for Practice for each component so that you can begin to think differently about this process.

First, not surprisingly, teachers felt that a visit by a mentor increased their content knowledge. In asking specific questions, both in their letters and during the visits, there was the expectation that the visit would lead to clarification and to confirm their expertise. Additionally, the response letters revealed enthusiasm for the strategies and suggestions developed through the post-visit conversation. Teachers enjoyed the opportunity to talk with a person outside of their school about their classroom and teaching. An outsider looks at their classroom through a different lens, seeing teaching and learning more objectively. Not being a stakeholder in the curriculum, an outside observer has the opportunity to look with broad strokes as well as see the finer points often missed by those immersed in the school and its programs. Many stated that they had never had the opportunity to discuss their classroom in this way.

We also believe that encouraging— indeed requiring—teachers to determine the content and context of their observation experiences was instrumental in creating deep and lasting learning. The teachers took on the challenge of extending their comfort zone and tried something new or difficult benefited from taking the risk. Patricia reflects,

> Rather than *telling* us what she wanted to see, she *offered* us the opportunity to decide what *we wanted* her to see. This was very foreign to me. My supervisor always told me what she wanted to see me do. I have since realized that teacher selection plays an important role in professional growth.

These aspects of the mentor visit experience led teachers to express feelings of reassurance and inspiration in their response letters. Teachers are not being held to a goal established by their supervisor, but instead set goals for their own personal and professional growth. Because there was space to ask questions without fear of judgment, an opportunity to talk over details of a lesson, and the experience of having someone notice and make explicit their strengths, the visit for some renewed and invigorated them. Karen writes,

> I have gotten over my fear of guided reading because I am more confident of what it should look and sound like. If I didn't give it a try with you I would not have known how to effectively teach both guided and strategy lessons. I also learned not to be scared to have someone observe something out of your comfort zone.

For some, the experience was transformational. Sara says, "To be honest, I might have just stopped interactive writing when I realized after 30 seconds that this was not working but I am happy you were there and it made me continue. It is only through working through it that I will become better." Fatima asked for help with her "difficult class" who she felt were unproductive during writing workshop. After the site visit and conversation she explains, "As it turns out, this was not a problem in terms of my students, but rather a problem with me" and goes on to reflect upon the changes she is in the process of making.

What kinds of classroom visits have you experienced? How do you prepare for a visit from a principal? From a literacy coach? From a colleague? Think about how to take charge of an observation by inviting a trusted colleague to watch you teach in an area for which you need more support. Consider the areas of growth you've identified in the survey (chapter one) as a place to start. Use the Tool for Practice *Letter of Invitation* to help you frame your letter of invitation for a classroom observation from a trusted colleague, perhaps someone you have identified as you build your learning community, as discussed in chapter three.

Viewing Videotapes and Giving Feedback to a Peer

Teachers can be very critical of their own practice. Anna writes,

> Although I am constantly assessing myself, I do not do it in a formal manner.
> I mull over my day on my way home or at night before I fall asleep. Actually, I
> tend to berate myself for all the things I could do better and vow to fix them.
> Watching myself on videotape allowed me to step outside of the situation and
> observe my own teaching. I saw things that I could definitely improve but I
> also saw some things that I am proud of. I realized that I don't admit that
> there are things I do well. This experience really helped me realize that I need
> to be more active in self-assessing myself!

Viewing videos of themselves also helped teachers to focus their attention on the children's engagement. They begin to see the connection between
teaching and learning as they embrace the opportunity to see their students as
active members of their classroom. New teachers are naturally very concerned
about their performance. Watching your own video allows you to see your
teaching from the children's perspective. Lana shared, "The video exchange
forced me to evaluate myself. The thought of watching myself on video fright-
ened me. But it really opened my eyes to see myself as my students see me."

Videotaping is a wonderful way to learn about your teaching. In the Tool
for Practice *Let's Go to the Videotape!*, begin to plan your videotape experience. Use this tool to record your pre-observation concerns and planning.
You can return to this form after you watch yourself. Teachers have found
it helpful to identify an area to focus on such as asking questions, or student
engagement, so as not to be overwhelmed by viewing the tape.

Giving feedback to a colleague or friend can feel uncomfortable. Helene
reflects, "The tough part about this was that I had to provide Martha with
effective feedback, and also give some suggestions or changes she could
make during her teaching. This was extremely challenging for me because I
was so concerned about hurting Martha's feelings or being out of line." But
working through this challenge to watch very carefully and find the right
words can strengthen your teaching expertise.

The peer videotape share can be a valuable professional development tool
that can help you recognize your teaching strengths as well as identify areas
that you would like to work on. Watching a colleague's lesson similarly helps
you identify strengths and strategies in teaching, as well as challenges you to
come up with effective suggestions. It can also confirm that your classroom
is not as unique as you might think, that others share the same challenges
and struggles in instruction. Here are some reflections from three teachers
who have watched their own lesson and provided feedback to others:

> I really valued the chance to watch and listen to myself interact with stu-
> dents. It made me very aware of important details such as tone and teacher
> language. I was able to see myself in a way I never was able to previously. I
> noticed things about myself as a teacher and reflected on best practices. I see

that this can be a very powerful professional development tool. Receiving feedback from a peer was extremely helpful because it allowed me to compare my own opinions/feelings with another professional's. Giving feedback to a colleague about their video lesson was also an opportunity to push myself professionally by practicing how to give suggestions in such a way that they are well received while at the same time pointing out what was effective and done well. (Ellen)

I truly enjoyed watching Sara's video, and having the opportunity to critique her lesson. It was great having the chance to observe another 6th grade teacher's classroom. When writing Sara's feedback on her reading conference, I was able to use the information that I learned from previous classes and my own teaching experience to give her feedback. I also thought it was helpful having her watch my lesson and receiving some great ideas from Megan on how to make my lesson better. I found it very helpful to read her comments on my lesson. As a teacher it truly helped me to know my strengths and weaknesses. (Linda)

Giving feedback to my peer did at first seem a bit tricky because the way something is worded can determine how it is taken to such a strong degree. However, when I was preparing to respond to my partner's lesson, I just thought about how I appreciate useful advice for my own growth as long as it is given respectfully and I tried to do the same. I would definitely like to use more videotaping as a means of reflection and professional development in the future. It made me so aware of details such as my tone and teacher language. (Rachel)

The Tool for Practice *Viewing a Colleague's Lesson* will help you as you observe another teacher's work. As you can see from the comments presented earlier, watching a lesson with the goal of supporting another teacher's practice is mutually rewarding. Talking to your colleague ahead of time helps you to establish a mutual understanding of the classroom context and your colleague's perspective on the lesson. By being a careful listener and thoughtful questioner, you can hone your abilities to think carefully about literacy instruction.

Take charge of the observation process so that you can learn. You may have a literacy coach in your building. Invite her or him to your classroom, to help you work through an area of discomfort. There may be another teacher on your grade level who you would like to visit. Find a time to exchange visits. It can also be helpful, as the teachers we work with report, to visit in a different grade level. Collaborate with a teacher or other school professional with whom you feel comfortable. Another idea is to use video to share your teaching with a friend from your preservice teaching. With digital video, it is easy to have a video exchange across the country!

Opening your doors through the observation process is scary but rewarding. As Diane states, "I have learned that sometimes you need to just throw yourself into a situation and not worry about the nerves you may be having. The best growth will come from situations that you are not very comfortable with but allow yourself to be a part of the process."

Tools for Practice: *Letter of Invitation*

Here is letter framework you can use to help you craft a letter to a trusted colleague inviting him or her to visit your classroom to help you learn to be a stronger literacy teacher. Fill in the blanks to start, then you can add more details to make the letter your own.

Dear _____,

I would like to invite you to visit my _____ to observe me _____
_____

As I think about my literacy teaching, I feel I would benefit from some support in _____. I will share some background of my thinking _____
_____
_____
_____.

When you visit my classroom you will see _____.
Here are some questions and concerns I have: _____
_____.

I believe you can assist me by _____
_____.

I look forward to your visit!

Tools for Practice: *Let's Go to the Videotape!*

"Do I really sound like that?" is a common first reaction to watching yourself teach on videotape. But once we can move beyond the discomfort of seeing and hearing ourselves, we can learn much about our teaching through videotaping our own practice (table 4.1). It helps to articulate our concerns, as you can do in the first column, and to reflect on what you see and think afterward, in the second column.

Tools for Practice: *Viewing a Colleague's Lesson*

Visiting a colleague's classroom is a wonderful way to deepen your understanding of literacy teaching. You may want to visit a teacher who has expertise in an area you would like to learn more about. You may, like the teachers we've worked with, want to exchange visits so that you can participate in a peer-coaching role. In both instances, it is critical to allow time to meet before and after the lesson.

This tool (table 4.2) is an open-ended form to help you structure your observations and peer coaching.

Table 4.1 Let's go to the videotape!

Concerns/Before the taping	Reflections/After viewing the tape
Anticipating the videotaping	
Lesson	
I hope to learn…	

Table 4.2 Viewing a colleague's lesson

Before the lesson	Notes during lesson	After the lesson: thoughts/questions/ideas
Lesson focus:		
Teaching focus:		
Student focus:		

Suggested Resources

Allen, J. (2006). *Becoming a literacy leader*. York, ME: Stenhouse.

Boushey, G., & Moser, J. (2006) *The daily five*. York, ME. Stenhouse.

Fountas, I., & Pinnell, G.S. (2001). *Guiding readers and writers: Teaching comprehension, genre, and content literacy*. Portsmouth, NH: Heinemann.

Miller, D. (2002). *Reading with meaning*. York, ME: Stenhouse.

References

Acheson, K., & Gall, M. (1987). *Techniques in the clinical supervision of teachers* (2nd ed.). New York: Longman.

Allen, J. (2006). *Becoming a literacy leader*. York, ME: Stenhouse.

Armstrong, A.J. (1999). Improving student teachers' reflection and personal practical knowledge. ED 459 146, Retrieved online April 3, 2009.

Casey, K. (2006). *Literacy coaching: The essentials*. Portsmouth, NH: Heinemann.

Darling-Hammond, L., & McLaughlin, M.W. (1995). Policies that support professional development in an era of reform. *Phi Delta Kappan, 76*(8), 597–604.

Dearman, C.C., & Alber, S.R. (2005). The changing face of education: Teachers cope with challenges through collaboration and reflective study. *The Reading Teacher, 58*(4), 634–640.

Dicks, M.J. (2005). Show me the way. *Educational Leadership, 63*(3), 78–80.

Frieberg, H.J., & Waxman, H.C. (1988). Alternative feedback approaches for improving student teachers' classroom instruction. *Journal of Teacher Education, 39*(4), 8–14.

Fullan, M. (2000). *Change forces: Probing the depths of educational reform*. Levittown, PA: Falmer.

Fullan, M., & Hargreaves, A. (1996). *What's worth fighting for in your school?* New York: Teachers College Press.

Gallimore, R., & Tharp, R. (1990). Teaching mind in society: Teaching, schooling, and literate discourse. In L.C. Moll (ed.), *Vygotsky and Education: Instructional implications and applications of sociohistorical psychology* (pp. 175–205). Cambridge: Cambridge University Press.

Greenwalt, K.A. (2008). Through the camera's eye: A phenomenological analysis of teacher subjectivity. *Teaching and Teacher Education, 24*, 387–399.

Henk, W.A., Moore, J.C., Marinak, B.A., & Tomasetti, B.W. (2000). A reading lesson observation framework for elementary teachers, principals, and literacy supervisors. *The Reading Teacher, 53*(5), 358–369.

Joyce, B., & Showers, B. (2002). *Student achievement through staff development* (3rd ed.). Alexandria, VA: Association for Supervision and Curriculum Development.

Lortie, D. (1975). *Schoolteacher*. Chicago: University of Chicago Press.

McDiarmid, G.W. (1999, April). *Tilting at webs of belief: Field experiences as a means of breaking with experience*. Paper presented at the annual meeting of the American Educational Research Association, San Francisco, CA.

Roehrig, A.D., Guidry, L.O., Bodur, Y., Guan, Q., Guo, Y., & Pop, M. (2008). Guided field observations: Variables related to pre-service teachers' knowledge about effective primary reading instruction. *Literacy Research and Instruction, 47*(2), 76–98.

Sabers, D.S., Cushing, K.S., & Berliner, D.C. (1991). Differences among teachers in a task characterized by simultaneity, multidimensionality, and immediacy. *American Educational Research Journal, 28*, 63–88.

Showers, B. (1985). Teachers coaching teachers. *Educational Leadership, 42*(7), 13–18.

Smith, D.J. (1996). Peer coaches' problems with videotape recording for teacher observation. *Action in Teacher Education, 17,* 18–28.

Song, K.H., & Catapano, S. (2008). Reflective professional development for urban teachers through videotaping and guided assessment. *Journal of In-Service Education, 34*(1), 75–95.

Swafford, J. (2003). Teachers supporting teachers through peer coaching. *Support for Learning: British Journal of Learning Support, 13*(2), 54–58.

Toll, C.A. (2005). *The literacy coach's survival guide: Essential questions and practical answers.* Newark, DE: International Reading Association.

U.S. Department of Education. (2001). *Improving the odds: A Report on Title 1 from the Independent Review Panel,* Washington, DC.

Valencia, S.W., Place, N.A., Martin, S.D., & Grossman, P.L. (2006). Curriculum materials for elementary reading: Shackles and scaffolds for four beginning teachers. *The Elementary School Journal, 107*(1), 93–120.

Vogt, M., & Shearer, B.A. (2007). *Reading specialists and literacy coaches in the real world.* New York: Pearson.

Walpole, S., & Blamey, K.L. (2008). Elementary literacy coaches: The reality of dual roles. *The Reading Teacher, 62*(3), 222–231.

5

Acting into New Ways of Thinking

Oh, no! I knew I was not good at expressing myself through movement and rhythm. I was very nervous and wanted to somehow disappear in the back of the room. But everyone was standing in the circle and I had no other choice but to join them... What does modern dance have to do with my teaching, anyway?

—*(Mary Ann, third grade teacher)*

Late and exhausted, after a full day's work with her third grade students and after being stranded on the subway platform waiting for a disabled train to be removed, Mary Ann finally arrived in her graduate class. Instead of finding an empty seat in the back of the room, she was taken aback as she saw chairs pushed against the walls, and her classmates standing in a circle taking turns clapping out the rhythm of their names.

Challenge and Opportunity

New teacher Mary Ann is not alone in her resistant stance. Her work with twenty-seven third graders in a high-poverty school requires her to obtain as many practical skills as possible to handle day-to-day life in the classroom, not how to learn through art. Most new teachers of the elementary school, like Mary Ann, enter the teaching profession after one semester of student teaching experience, which is divided between a lower grade and an upper grade classroom. Prior to student teaching, they have opportunities to observe in classrooms while taking the foundations and subject specific methods courses. As they step into the classroom as neophytes, what they've learned in the teacher preparation programs is often tempered by the complexities and constraints of school realities. Given the immediate assumption of responsibilities as the teacher, new teachers often adopt a survival style in the induction years of teaching and do what works. They often follow the teachers' guides lesson by lesson not confident of their own abilities to make instructional decisions, afraid they may miss something. Scope and sequence charts, rather than their knowledge of the curriculum in general and the children in their classrooms, guide their instruction. They tend to have a "tell me what to do tomorrow" approach to professional development. They want to gather as much How-tos as possible to help them plan and implement curriculum, organize the classroom environment, manage student learning individually as well as in small groups, to name a few.

However, there is nothing inherently wrong with the initial survival style. The new teacher adopts this style because he or she feels the need to survive. The challenge is, how can the new teacher develop the potential to be more deliberate and thoughtful without letting the survival tactics become engrained habits? What kind of professional development could help the new teacher develop an understanding of children and their learning, content knowledge, and pedagogical skills that lead the neophyte to become a stronger teacher over time?

All through the initial teacher preparation program, including the student teaching experience, as well as the first years of teaching, new teachers are focused on their teaching and the curriculum to preclude serious reflection and deliberation (Kagan, 1992). Their emergent teacher expertise has been developed from the perspective of teachers. Staff development efforts and graduate courses need to provide new teachers with a context and climate for learning where they learn to examine their histories and assumptions about teaching and learning. Opportunities that allow new teachers to walk the walk and talk the talk as a learner can help to bridge their personal knowledge as learners and professional knowledge as teachers. Activities that require new teachers to confront their personal knowledge can be helpful in moving them toward its critique and reconstruction, and thereby foster their growth toward a greater capacity for deliberation.

In this chapter, you will get to know a few teachers and learn what it was like for them to be engaged in learning opportunities as learners themselves, and how this learning impacted their teaching. You will see that they often have to take a calculated risk in order to try out activities or strategies, which then afford them with fertile ground for new perspectives. You will be introduced to strategies that might intrigue you to put yourself in the learner's shoes and experiment how this experience can help you become more deliberative in teaching.

Stories from the Field

Second-year teacher Mary Ann had enrolled in our graduate program that began two weeks before she met her third graders for the first time. She felt that she was learning to teach in a trial by fire fashion. Juggling her first-year teaching and graduate school was onerous and taxing, yet she credited her graduate school courses and classmates for sustaining her through the tough year. As she began her second year of teaching, she was eager to add new items in her teaching "tool box." But in her Literacy Assessment course, she was puzzled by a portfolio unit focusing on studying a modern dance. She felt both bewildered and impatient with the experiential activities that were meant for the course participants to find their entry points into the work of art. "My school expects me to walk in there everyday and do a good job teaching my students. What does modern dance have to do with my teaching, anyway?" she wondered.

Learning in the arts also brought back unpleasant memories Mary Ann had as a child in her elementary school art classes. She remembered that she always "made a mess" in art classes, and that her work would either be ruined or be hung up in inconspicuous places. She claimed that she didn't have creative talents and brought this perception with her every time she encountered arts. Her graduate class was no exception. She was hesitant first when the class shared their collages gallery style. Even though she was relieved knowing that she didn't have to put her name on her collage, she still tried to find a less conspicuous spot to display her work in the gallery. The more she looked at her peers' work, the more she felt inadequate. She wrote:

> I guess one good thing about the workshops was that they allowed me to get to know a different aspect of my classmates. It was interesting to see everyone's work on the wall. I could see that some of them were really talented. But the more I looked, the worse I felt about my own work, which was sadly tucked away in a corner.

Joyce felt the same anxiety as Mary Ann when she knew that she would need to share her personal narrative draft with a peer response group in the writing methods course. She was not as reluctant to keep a writer's notebook and to write alongside the children in her own classroom. She thought that it was a good idea to go through the process of drafting, conferencing, revising, and editing process, as she understood that the personal narrative writing project, as part of the course work, was conducive to experience the writer's work in collecting ideas, uncovering and discovering the significance of her own writing, but she was not thrilled about sharing her writing with others. In fact, she was terrified.

Like Mary Ann, Joyce's fear of sharing her work stemmed from her writing experience in the middle school. She had vivid memories of how she labored over her composition each week. But most of the time, her papers were returned bleeding with the remarks of her teacher's red pen, "Run-on!" "Sentence fragment!" "Awkward!" "Vague!" She would then go over each red-pen remark, correct each error, and carefully copy over the entire paper. She even believed that her ideas and content had no merit since her teacher never looked beyond the errors.

Joyce was working on the edge of her comfort zone as she wrote her personal narrative. She faced two road blocks: memories of those red-pen marks, and the overwhelming thought of sharing her work with her peers. Therefore, her writing process was constantly jeopardized as she was grappling with both form and content. She was afraid that her syntax was not good enough. She did not feel that her story was important enough. But since this was required course work, she had not choice but to push forward.

You will learn later in this chapter that both Mary Ann and Joyce benefited from these learning to learn experiences. But the questions still remain, does their action lead to new insights about their learning processes? Does their work as teacher benefit from their learning? Will their own learning, as well as their new insights about themselves as learners affect their teaching? If so, to what degree?

Research and Reflection

The idea of putting teachers in the learners' shoes is a powerful one. Research suggests that those who know themselves as learners and are able to monitor and change their learning strategies accordingly are better able to transfer their learning to new contexts (Bransford et al., 2000). Studies of multiple intelligence theory in teacher education affirmed the value of student reflection in building self-confidence and learning-to-learn skills (Kallenbach & Viens, 2004). As is shown in the Strategies and Tools section of chapter three, teacher reflection tends to fall on three levels, Implementer Reflection, Observer Reflection, and Transformer Reflection. Authentic learning opportunities such as learning through the arts, being in book clubs, or writing for real world purposes and audiences remove the possibilities to reflect on the first two levels, thereby leading the teacher/learner to reflect on his or her own learning processes.

Learning and knowing do not happen in isolation in an individual's mind, but are fundamentally social as the learner interacts with others in contexts that value certain learning modalities. From teacher education methods courses to field practices, teachers learn through experimentation, discovery, and application of content knowledge and procedure knowledge. In authentic learning tasks, teachers must generate and construct knowledge. Knowledge, then, is not viewed as "a thing to be possessed but as activity to be engaged in" (Gill, 1993, p. 68). Wells (1999) defines knowing as "the intentional activity of individuals who, as member of a community, make use of and produce representations in the collaborative attempt to better understand and transform their shared world" (p. 76). These studies support the learning activities that are social, and that allow the learner to show what they know through their own meaning-making processes, including self-reflection.

Building on Fuller's concerns based model of teacher development (1969), which includes three stages that the teacher trainees progress through—self, task, and impact—Conway and Clark (2003) found that beginning teachers' concerns shifted outward from self, to tasks, to students, as well as inward, from those about personal capacity to manage their classrooms to concerns about their personal capacity to grow as teachers and people. Authentic learning tasks allow the teacher/learner to discover and define himself/herself under a new light, through this inward journey in search of his/her own capacity for learning, to transform and redefine his/her own perceptions of learning and teaching.

It is well established that the teachers' thinking affect their actions (Clark and Peterson, 1986). In a review of literature, Hall (2005) discussed several studies that focused on learning experiences of teachers and how these experiences led to better understanding and more frequent implementation. The findings suggest that it is through action, not thinking, that teachers can gain new ways of thinking that in turn guides their action in the classroom. Intellectual ideas, concepts, and information are important, but merely studying them is not likely to change one's behavior. As teachers in the classrooms, we do pass on a lot of information. But the information is only communicated by those who have demonstrated their own capacity to be able to do exactly what they are teaching.

It is found that those who know the ways they learn best, and perhaps the way they teach best, are better able to transfer that knowledge to new contexts (Bransford et al., 2000). Though Bransford's study was on learning strategies, it is not hard to speculate that those learners who personally benefit from learning strategies are more likely to integrate innovative methods into their own instructional practices.

Revisit the Story

As Mary Ann listened to her peers' comments on her collage, her fear and resistance gradually turned into fascination. She realized that the "gallery" was not set up to evaluate everyone's work. Instead, the teaching artist encouraged the participants to notice and describe what they saw. Some noticed that Mary Ann's collage had a lot of sharp angles and edges, as if to express energy or anger; others commented that the blue circles hanging off the edge of her work seemed to be peaceful and somewhat uncertain; still others argued that it was a work of contrast, that it was eye-catching because of the contrast. These discoveries of hers and others' work surprised Mary Ann. She realized that the whole process of making the collage was not about doing it the right way or having creative talents. She came to understand that her creative process was one that invoked responses, and that she had contributed to the construction of meaning and understanding of her learning community. She began to examine her own assumptions about herself as a learner. She wrote:

> I have always thought of myself as not very creative, but by making my collage, and learning how others perceive my work, I was able to prove myself wrong. Making that collage has helped to bring out a strength that I never really thought was there. I am slowly learning that learning doesn't have to be right or wrong.

Her words illustrate that Mary Ann gained new insights about her own perception of her capacities. She found strength in knowing that she was capable of creating meaning using art forms and that her meaning-making process was situated within her own lived experience, not from artistic talent that she had felt was beyond her reach. It is important to note that it was in the social learning environment, through the responses of her peers to her work, that she gained these new insights. As a teacher, you may be able to relate Mary Ann's learning with many of your children's learning—that when you respond to their stories and say, "Oh, I didn't know that when you shot that ball through the hoop, you changed those boys from enemy to friends! I can see why this was such an important story for you to tell!" Then all of a sudden, your young writer's eyes lit up and he exclaim, "Yeah, I am going to add here that before that ball went in, those boys didn't really like me and they made fun of me!" All he needed was to know that his story had meaning to a reader, and he was going to make it even more meaningful.

As she looked back, Mary Ann realized that she had resisted the experiential workshops associated with the modern dance every almost step of the way. Her portfolio reflection detailed how each of the workshop started with great resistance. In her reflection of the first two workshops, Mary Ann repeatedly used words such as "uncomfortable" and "confused." When she described her learning during the workshop after the performance, she had a new revelation about the cause of her discomfort—that she was always "following the rules" that had given her a sense of order and accomplishment. But she realized that following the rules had prevented her creative and innovative ideas from flowing; hence she was very resistant to open-ended tasks. In the end she wrote:

> The most important insight of this portfolio project for me is the realization that some students may feel as uncomfortable with reading/writing projects as I felt with these artistic experiences. The question I have to ask is this: how can I lead that student to self-discovery, empowerment that will help him/her to grow in a way that maybe I'm only beginning to realize as an adult?

Mary Ann's insights illustrate that learning could be more profound when one works at the edge of one's comfort zones. The active engagement in art activities, dialogue with the others, and sharing and constructing knowledge together gave rise to insights similar to Mary Ann's. Walking in the shoes as learners themselves allowed Mary Ann and her peers to consider the learning of their students. Many questioned the pressure they put on their students to perform or produce with little or no choice of their own. Here is a sampling of the thoughtful questions the course participants asked about their teaching and their students' learning:

- Did I give my students choices? Was there scope to be creative?
- Could [my students] take risks?

- Did [my students] learn anything beyond boundaries set by their teacher?
- Did I give my students enough time to explore and come to conclusions themselves?
- Did I allow my students opportunities to explore other possibilities and explanations?
- Do [my students] feel connected to the work that they are doing, like I'm connected to mine?

Joyce went into her peer-response group with trepidation. But after she finished reading her narrative, comments and questions from her peers made her feel validated, supported, and understood. Her peers were mesmerized by her childhood memories of Trinidad. "Fascinating," they commented. "Tell us more," they requested. Her professor told her in a conference how she had admired her courage to not just adapt to the lives in the new world, after being uprooted from her hometown in Trinidad, but to reinvent herself. Never in Joyce's writing life before did she feel not just a sense of relief after her writing reached an audience, but a sense of eagerness bordering on enthusiasm that urged her to go back and revise. She wrote:

Unlike the dreaded red pencil remarks I used to receive in middle school, the responses I was given inspired me to want to revise. My response group wanted to know more about my life in Trinidad. Professor H said that if mine was a turning point story, she needed to see turning from what to what. I learned a new word from her—juxtapose. I would need to contrast my lives in two different worlds so that I can show my reader how I was grappling with coming to a new world, rather then telling them over and over that it was difficult and confusing.

"Show my reader." It is hard to believe that these words were uttered by the same person who had been agonizing about sharing her work not too long ago. Acting as a writer in a supportive community of writers not only gave Joyce confidence, but a sense of audience and purposes. With new-found confidence as a writer, Joyce revised with rigor, and even surprised herself with how meaningful her story turned out to be.

Subsequent course work required Joyce to design a strategy lesson in revision and teach it to her class. With her own revision process fresh in her mind, and a changed perception of herself as a writer, Joyce set to work with a vengeance. Why wouldn't she? The idea of helping her students to revise without the red pen invigorated her. If she herself felt so empowered by her peers' and professor's responses, why would she not do the same with her own students? As she planned the lesson, she wished that her middle school teachers could show her strategies she could actually use to make her writing better instead of using the dreaded red pen to catch every error she had made. Even though she dutifully corrected those mistakes, she still hadn't learned how to revise her own work until now. Now that she was a teacher, she knew that she was going to teach her students in a way that she wished she had been taught.

Later, Joyce reported that her strategy lesson on revising for coherence was a success. Her lesson idea was based on her observation of her students' writing. Many of her students' thoughts tend to be marooned in each sentence, rather than writing several sentences in a passage to develop an idea. She decided to help students to revise for coherence by paying attention to creating thematic unity in the passage. With the newfound strategy, Joyce taught strategy lessons and conducted conferences, using her "before" and "after" passages to explicitly show her students how to revise for coherence. She wrote:

> This writing experience has helped me in relating to my students. By putting myself in their shoes I was able to observe the challenges they might encounter during a writing cycle. I used my writing and revising as examples when I teach writing. My students not only felt empowered because they feel connected to how grown-ups write and revise, they also saw explicit strategies to revise and craft and they were eager to try them out, just as I did.

The nature of the experiential learning opportunities such as those Mary Ann and Joyce found themselves in, similar to that of inquiry, is open, critical, and dialogic. The work of art and the act of crafting a personal narrative that is meaningful to the writer invite the learners to take a stance, to wonder, to ask questions, and to collaborate with others in seeking answers and creating meaning. Mary Ann's and Joyce's new insights about themselves as learners didn't come by simply by reading about how learners learn or by thinking about who they were as learners. You can't think your way into a new way of acting. Ideas and strategies are best assimilated and understood in the context of life applications. When you, the teacher, decide to put yourself in the learner's shoes, you certainly can act your way into a new way of thinking.

Strategies and Tools

I learned to swim as an adult, having grown up in one of the most inland cities in the world. Instead of having swimming lessons from instructors in the pool, I learned to swim in the ponds and bay of Nantucket Island from friends whose swimming histories were longer than my entire life. My learning was unstructured. There were no flotation devices or practice sessions, just my friends and I out to enjoy the delicious embrace of the water. I didn't choose to learn to swim. But just being on Nantucket for the entire summer made the opportunity to learn to swim too good to refuse.

Being in the water brought back memories of childhood play in the pool. I enjoyed playing in the pool, but never signed up for any lessons, because I got cold easily and the idea of taking classes even when you were cold didn't seem to be quite appealing. As an adult learning to swim, I was very aware of my own learning processes, my limitations as well as preferences. I knew that my toughest challenge was an innate fear of water. I never felt secure when my feet could not touch the bottom. I also had an infinite preference for swimming in the bay, where the calm waves and salt water made me lighter and freer, where I felt stronger as a swimmer.

At the core of acting into new ways of thinking is experiential learning. It brings learners into the learning processes through hands-on investigations or creations that actively engage the learner in perception, exploration, research, and reflection. According to adult learning theories (Brookfield, 1995), effective adult learning involves the following features, which need to be incorporated into your learning:

- giving learning a purpose;
- incorporating self-reflection;
- facilitating self-directed learning;
- including self-evaluation in assessment;
- valuing learners' experiences in instruction.

To ensure ownership of the learning process, your first consideration is choosing what it is that you want to learn. Then you must choose mentors and tools, as well as time commitment. You will also need to explore your previous experiences to see if you can tap into any prior knowledge. Table 5.1 serves as an example to a variety of authentic learning opportunities.

Keep a Learning Log

As discussed in chapter three, being aware of your own learning allows you to become more sure-footed. A learning log may provide you with a space where you can document and articulate your own learning process. You may adopt the double-entry format, as discussed in chapter three, which allows you to reflect on your learning. The key is to allow the process be the main focus. You may also document your affect as a learner, not just the actions you took. Articulate your resistance or preferences, when you notice them. Describe your discomfort and keep track to see if it gets better over time. Most importantly, document any new insight you might have,

Table 5.1   Interest inventory

Skills	Arts and crafts	Sports	Music and movement
First aid	Knitting	Wii sports game	Dance
Sign language	Scrap-booking	Kayaking	Yoga
A second language	Painting	Skiing	Guitar
Writing	Pottery making	Tai Chi	Singing

about your own learning, about the activities you were involved in, about your mentor or the mentoring processes.

Examine Your Own Histories

Our own experience, prior knowledge, as well as preconceived notions about ourselves as learners can help or hinder our learning. When they are helpful, you can use them as resources. But when they hinder our learning, they become baggage that could bog us down. Our preconceived notions are often products of our past learning experiences. So it is important to examine your own histories. You can do this mentally or in your learning log by asking yourself questions. For example, if you decide to study a second language, you might ask yourself:

- What memories do I have about learning a second language?
- Did I choose to learn this language?
- Did I have significant mentors who taught me a second language?
- Did I have any significant memories about learning and using the second language?
- How did I perceive people who speak this language? How did I perceive the country where this language is spoken?
- How did I learn? For how long?
- Do I know any other languages besides my primary language?
- What did I think about learning a second language?
- What positive experience did I have learning a second language?
- What negative experience did I have learning a second language?

Reflect on Your Learning Frequently

You can use the learning log as one container for both documentation and reflection. Just walking the walk as a learner is not enough. One must also talk the talk as a learner. Frequent reflection allows you to discover and define your learning, and question any preconceived notions you might have. You might ask yourself:

- What did I do today that was significant?
- Why was it significant?
- What was difficult/easy/challenging?
- What were my limitations?
- What have I learned about myself as a learner (or writer, dancer, painter)?

Questions such as these can help you learn with an awareness of your own learning. You will soon find that the ability to articulate one's learning process, to identify patterns and significance of the learning tasks can ultimately help you relate to your students. Don't feel guilty about taking the time to learn something that might not be considered survival-related or job-related. After all, your own learning process, new insights about yourself as a learner, new perspectives on the learning process itself can be shared with your students to inspire and to show an explicit model of authentic learning. Many Ann's reflection about her own portfolio process is a case in point. She wrote:

> Being asked to constantly reflect on our learning made deep perceptions unavoidable...Actually engaging in a reflective experience gave me the motivation to use [the portfolio] with my class. I found that having choice in my final outcome, and thinking about the processes helped me to do my best work. I feel, as a learner, that this type of assessment gave me more of an opportunity to demonstrate my learning and understanding of the subject under investigation.

Dust off your brother's guitar. Sign up for Tai Chi. Get the box of yarn and needles from the back of the closet. You will experience the joy of being a learner again. And your students will enjoy knowing you as a learner— underneath the teacher clothes you donned on—and feeling that they are learning alongside you.

Suggested Resources

Bransford, J., Brown, A., & Cocking R. (Eds.). (2000). *How people learn*. Washington, DC: National Academy Press.

Eisener, E. (2003). The arts and the creation of mind. *Language Arts, 80*(5), 340–344.

Gardner, H. (1982). *Art, mind and brain: A cognitive approach to creativity*. New York: Basic Books.

Gardner, H., & Perkins, D.N. (1989). *Art, mind and education: Research from project zero*. Urbana, Il: University of Illinois Press.

Greene, M. (2001). *Variations on a blue guitar: The Lincoln Center Institute lectures on aesthetic education*. New York: Teachers College Press.

Lave, J., & Wenger, E. (1991). *Situated learning: Legitimate peripheral participation*. Cambridge, UK: Cambridge University Press.

Levine, S.L. (1989). *Promoting adult growth in schools: The promise of professional development*. Boston: Allyn & Bacon.

Perkins, D. (1991). Educating for insight. *Educational Leadership, 49*(2), 5–8.

References

Bransford, J., Brown, A., & Cocking R. (Eds.). (2000). *How people learn*. Washington, DC: National Academy Press.

Brookfield, S. (1995). *Becoming a critically reflective teacher*. San Francisco, CA: Jossey Bass.

Clark, C.M., & Peterson, P.L. (1986). Teachers thought processes. In M. Wittrock (ed.), *Handbook on research on teaching* (pp. 255–296). New York: Macmillan Publishing Company.

Conway, P.F., & Clark, C.M. (2003). The journey inward and outward: A re-examination of Fuller's concern based model of teacher development. *Teaching and Teacher Education, 19*, 465–482.

Fuller, F.F. (1969). Concerns of teachers: a developmental conceptualization. American *Educational Research Journal, 6*, 207–226.

Gill, J.H. (1993). *Learning to learn: Toward a philosophy of education.* Atlantic Highlands, NJ: Humanities Press.

Hall, L. (2005). Teachers and content area reading: Attitudes, beliefs, and change. *Teaching and Teacher Education, 21*(4), 403–414.

Kagan, D. (1992). Professional growth among preservice and beginning teachers. *Review of Educational Research, 62*(2), 129–169.

Kallenbach, S., & Viens, J. (2004).Open to interpretation: MI theory and adult literacy learning. *Teachers College Record, 106*(1), 58–66.

Wells, G. (1999). *Dialogic inquiry: Towards a sociocultural practice and theory of education.* New York: Cambridge University Press.

Part II

Tensions of Literacy Teaching

6

Understanding and Moving beyond Labels

I don't think I can help him. He's all over the place. I think he's Special Ed.

*—(Carla, after her initial tutoring session with
seven-year-old Eduardo)*

Carla was a second-year fourth grade teacher enrolled in our MS in Literacy Program. She has been enjoying her work with older learners in a predominately monolingual classroom. At her school, she felt she had a handle on her fourth grade curriculum and was especially proud of her ability to prepare her students for their state exams. Yet Carla was anxious about her responsibility to tutor a struggling reader in her tutoring practicum course. Almost sheepishly she admitted, "I can spend a day with 28 fourth graders. But I am slightly terrified of working one to one with a little boy!"

Carla was assigned a bilingual first grade boy Eduardo, who had been referred to the program by his classroom teacher. As part of her plan for the first session, Carla read *The Hungry Caterpillar* to Eduardo and asked him to retell the story. Although he listened attentively, he did not respond to Carla's request. Carla then decided to assess his knowledge of letter-sound correspondence using magnetic letters. Filling the table with letters, she asked Eduardo to match letters to the sounds she made. Bewildered by the sea of letters and by the confusing directions, Eduardo again did not respond. At this point Carla came to me, slightly embarrassed, somewhat exasperated, and claimed that Eduardo was a special education child. Adamant that she could not, would not be able to teach him, she wanted to know if she could switch children so that Eduardo could work with a teacher who had a special education background.

Challenge and Opportunity

What did Carla see that might have led her to say "Special Education"? Carla's story reminds us that we may use an educational label before we fully attempt to describe and understand a child. Additionally, Carla did not recognize attributes of an English Language Learner (ELL) in Eduardo's literacy behavior, contributing to her initial mislabel of Special Ed. How might Carla move beyond a label and find a way to work with Eduardo?

Carla's use of the label "Special Ed.," based on one meeting with a seven-year-old boy, reveals both her initial assessment of how Eduardo's literacy behavior did not match her understanding of where he should be as well as, and as importantly, her sense of being overwhelmed by her responsibility to teach him. Labeling him as "Special Ed." was one way for her to say, "I can't handle this. I'm not feeling comfortable."

When you hear a student called "Special Ed." what do you think? What about hearing a student called "gifted" or "hyper" or "a problem?" Do those words raise concerns? As teachers we may think we share an understanding about what these labels mean, but labels can actually trigger stereotyped alarms that impede further growth or refined understanding of the child's specific literacy learning needs.

In chapter two we explored how our "teacher" language impacted the children we teach. In this chapter we unpack and analyze the educational terminology that is used to categorize and classify the children in our classrooms. How do terms such as "learning disabled" or "English language learner" or "gifted" help us understand and differentiate instruction for children? When and how do these labels prevent us from fully understanding the learning needs of the individual children? When do these labels allow us to back off from our responsibility to teach *all* children to the best of our professional abilities? This chapter unpacks the dangers of pasting educational labels on children rather than assessing and addressing the very real literacy needs of the diverse children we teach.

Stories from the Field

"He's read all of the Harry Potters! What can I teach him?" Understanding Gifted Learners

A peek into Rosa's second grade classroom library corner reveals many book bins—mysteries, biographies, sports. There are also books classified by Fountas and Pinnell levels, from A to Z. Yes, Z. Rosa had to borrow books to fill the bins of the end of the alphabet, as she has never had a second grade student in her class like Oscar. When her colleagues at P.S. 5, a mid-sized public elementary school, nestled amid a public housing complex in a large city, saw Oscar's name on her register, they chuckled. "You'll have your hands full with him!" they warned. "He's too smart for us," her first grade colleague said.

With those words in mind, Rosa was both anxious and eager to meet Oscar. How hard can it be to teach a gifted reader, Rosa thought. On the first day of second grade, Oscar marched into the classroom, with a large hardcover book tucked under his arm. "Hey!" he announced loudly, as children were putting backpacks into the closet and settling into their seats, "Hey, everyone. I've finished this Harry Potter book and Dumbledore dies!" Rosa internally winces, while her new second graders groan, "Oscar!"

As September moves on, Oscar continues to perplex Rosa. In her initial reading assessment, she finds that indeed Oscar does read independently at a level T, far beyond her other second graders. He is a precocious and voracious reader. Finishing quickly, then bored, Oscar also begins to act out. The other children at his table complain that he is annoying and that he criticizes them too quickly. Rosa's first thought is to give him extra work but as she's not exactly sure what to give him, it's more of the same. In October Rosa is ready to begin guided reading groups and reading partners but doesn't know what do with Oscar. It is only October, she says, and I really don't know what to do.

Rosa appreciates Oscar's talents, but is unsure as to how best to teach him. He's a different kid than she's used to and it is hard for her to figure out how to address his needs along with those of the rest of the class. Oscar's strengths are also causing him problems; he continues to be excluded by other kids who don't get his Harry Potter references and don't like his criticisms. Rosa wonders how she can both academically challenge Oscar and help him become part of the class.

"What should I work on first?" Understanding English Language Learners

Janet is a new teacher working in an urban public elementary school. She is teaching fourth grade and most of the children in her class are ELLs. Her school has implemented an extended day program where many of the children receive additional instructional time after the regular school day ends. One of the goals of the after school program is to ready the children for the upcoming standardized tests. Janet articulates her experience of teaching as she considers her experience tutoring Karina, a fourth grade English Language Learner:

> As I continued to work with her I realized that her difficulty with the English language was still creating a lot of academic challenges for Karina. I realized that although I was expected to focus on test prep, Karina had more basic needs to be addressed. Working with Karina made me think about struggling students and how we force them to participate in instruction that is above their level. At first I tried to work with Karina on test prep activities but found that she lost interest and her real needs were not being addressed. So I stopped doing test prep activities and started with activities that met her needs. I think I was able to make more progress with these activities! This also made me think of how I work with students in my own classroom and how I really need to think about what I am teaching and if it is academically right for my students or not.

Janet is struggling to figure out how to better understand and meet the literacy needs of Karina, an ELL, which causes her to reconsider her instruction in her classroom. As you think over your own classroom, are there students for whom English is a second or third language? What questions does this raise for you?

Research and Reflection

As the stories from the field reveal, there is a great diversity in our classrooms. Think about your own students. There may be students with Individual Educational Plans (IEPs), students who receive English as Second Language (ESL) services, students who are achieving way above the grade standards, and children who seem to be struggling to reach your grade standards. These may be the children who challenge you and, despite your professional stance, frustrate you as well. They push you out of your comfort zone.

Recognizing, thoughtfully assessing, and finally appropriately developing instructional plans to meet all children's needs is a tall order. As a new teacher it may seem overwhelming. Tomlinson et al. (1994) found that while pre-service teacher education programs help novice teachers recognize the importance of recognizing, assessing, and differentiating instruction to meet the needs of diverse learners, it is very difficult for new teachers to put those beliefs into action. Moon et al. (1999) concurred, stating," The complexity of recognizing, understanding, and addressing those needs can be overwhelming for early teachers who may lack some or many of the fundamental skills of teaching" (p. 56).

With the variety of efforts to raise standards of literacy achievement, and to meet the needs of a range of learners, it is vital we understand how to recognize and address the needs of all the learners in our classrooms. Even though many literacy instructional models or programs refer to adaptations or accommodations for a range of learners, in practice teachers need to do more than follow the teacher's guide in order to fully reach all the learners in their classrooms.

In this section we review research that will help us unpack and understand the labels ELL, Gifted, and children who need more intensive support in reading in order to be better able to understand the specific strengths and needs those children bring to our classrooms. We believe that it is important for teachers to understand characteristics of these learners so that we can see the children in our classrooms more carefully rather than rely on stereotypes from labels.

English Language Learners

By the year 2050 nearly 40 percent of the children who attend U.S. schools will arrive speaking a language other than English (Lindholm-Leary, 2000). There are likely to be ELLs in all schools, urban, suburban, and rural. However, their teachers may primarily be monolingual and so unaware of what it is like to speak and learn in more than one language (Sleeter, 2001). Attention to the strengths and needs of ELLs in teacher education programs is minimal. Thus it is critical to examine our beliefs about ELLs in order to move toward a fuller understanding of the ELLs in schools. Moss and Puma (1995) found that ELLs were judged by their teachers to have lower academic abilities than their monolingual peers even though scores on independent tests revealed no significant differences.

Along with the increase in numbers, there is a great diversity in who is an ELL. Freeman and Freeman (2007) describe three groups of ELLs: children newly arrived in the United States with adequate schooling, children newly arrived with limited schooling, and long-term English learners. The needs of each group of children are different. A newly arrived child with schooling may soon catch up with her peers, while a child without formal education will need additional support. ELLs who have been in the United States longer than seven years may have had inconsistent instruction and require closer attention. It is essential that teachers know as much as possible about the ELLs in their classrooms—including first language, earlier educational experiences, and family concerns.

Krashen (2003) provides a useful model to understand the process of English language learning. According to Krashen, in learning a second language, understanding precedes speaking, production (speaking or writing) develops in stages, and the goals of instruction should be placed in the context of meaningful communication. When students are first learning a new language there may be a silent period, a time when children may understand much more than they are able to speak. Learning English can take time, generally up to seven years, for students to gain academic proficiency. The students' needs, interests, and feelings are important resources to draw upon throughout the process.

Instructional approaches that support ELL need to take these issues into account. In particular ELLs benefit from specific instructional strategies. Freeman and Freeman (2007), adapted in table 6.1, offer many ideas. Also, table 6.1 provides you with some strategies to try in your classroom.

Gifted Learners

A gifted learner, according to Renzulli (1998), demonstrates three characteristics: above average ability demonstrated on standardized tests and performance assessments; ability to be fully committed to tasks; and creativity. These characteristics manifest themselves in gifted readers as early reading ability, learning to read independently, longer reading stamina, and, at times, a desire to read a wide variety of genres.

Researchers (Gaug, 1984; Townsend, 1996) have looked at two different strategies to address gifted readers' needs in the classroom. An enrichment approach provides "learning activities providing depth and breadth to regular teaching according to the child's abilities and needs"

Table 6.1   Strategies for teaching English language learners

Strategy	Rationale
Print filled environment with labels in both languages	Multiple and varied exposures to print, in both languages, fosters literacy development
Provide culturally relevant and engaging authentic texts	Opportunities to relate to the contexts and people in texts foster comprehension and enjoyment
Content-rich thematic units	Students get both language and content; language is kept in its natural context; students have reasons to use language for real purposes; students learn the academic language of the content area
Variety of opportunities for oral language, reading, and writing in both languages	Allowing students to speak, read, and write in their first language supports the transfer of skills and strategies and builds self-confidence and pleasure in literacy activities
Predictable, clear, and consistent instructions and routines	When routines are clear and consistent, students are able to focus on the content of the instruction
Instruction that builds up students' knowledge and skills in their native language	Proficiency in students' native language supports academic achievement as students' develop English language abilities
Focus on similarities and differences between English and native language	English vocabulary is key in reading comprehension. If the first language has cognates that are similar, it can help in transferring that understanding
Frequent feedback	ELLs are supported, linguistically and socially, by feedback that supports their involvement

(Townsend, 1996, p. 362). Some benefits to enrichment are the opportunity to meet the needs of a gifted child within the classroom without explicit labeling and forestalling problems with a gifted child being frustrated or bored in a classroom. Problems, however, exist as well. The extra work may not be substantially different or the work may be "filler" or "busy work" that does not related to the curriculum. Acceleration "occurs when children are exposed to new content at an earlier age than other children or when they cover the same content in less time" (p. 361). An advantage to an accelerated approach is that there can be greater intellectual challenge, and some researchers (Guag, 1984) have shown that children in accelerated programs excel in future school experiences. Yet moving children ahead may not be appropriate for the social and emotional needs of children, the materials used in advanced curriculum may not relate to their interests, and it can be disrupting to the child's school experience (Vosslamber, 2002).

While it may appear that gifted children are "easy" to teach, because they may catch on quickly, it is clear that gifted children benefit from targeted instruction as well as other children. Moore (2005) suggests several instructional approaches to support gifted learners in their literacy development (table 6.2).

Table 6.2   Strategies for teaching gifted learners

Strategy	Rationale
Purposely select texts with challenging and unique language structure and point of view, ambiguity, rich and varied vocabulary	Increasingly complex and engaging texts support and expand gifted readers' comprehension and vocabulary development
Introduce texts that have gifted characters as protagonists	Being able to identify with characters is important for gifted readers' development of social and emotional identities
Provide opportunities for inquiry reading that extends the classroom readings	Inquiry projects acknowledge and affirm gifted readers' unique strengths and interests, helping them to maintain interest and enthusiasm in school
Incorporate instruction in critical reading strategies	Although gifted readers may read advanced texts, they need instruction in how to critically read and analyze texts
Ask and encourage high level questioning	High level questions challenge readers to deeply engage with the text
Integrate opportunities for student-generated reading and writing	Within class units and themes, providing choice and opportunities for in-depth projects will challenge gifted readers

Special Education Learners

There are a wide variety of special needs encompassed under the label of Special Education. Some children may have an IEP and be in a general education classroom. Other may need a specialized classroom setting. Historically, a reading disability definition had three concepts: (1) children with a reading disability were achieving below their cognitive potential; (2) a reading disability is caused by a biological deficit; (3) need to rule out other causes such as mental retardation (Spear-Swerling & Sternberg, 1996). With this definition came a division in educational responsibilities. According to Caldwell and Leslie (2009), "Special educators taught children with learning disabilities and remedial reading teachers worked with children who were not placed in that category. Both groups employed different assessment tools and instructional techniques" (p. 2). However, researchers such as Allington (2006) have found these criteria ineffective, arguing that those children then frequently receive less effective reading instruction in schools. Dudley-Marling and Rhodes (1996) identify programs in schools that are either specialized, and remove the struggling reader from the classroom or eclectic approaches that try a bit of everything. They argue that for many students with reading disabilities what are most needed are "frequent, intensive, explicit, and individualized support and direction."

Response to intervention (RTI) is a new process to determine learning disabilities that moves away from the historical definition noted earlier. RTI originated in the President's Commission on Excellence in Special Education

(2002). This report argued that children with learning disabilities should first be considered general education students and that special education should use a model of preventing failure, rather than waiting for children to fail before they received academic support. Instead of using the discrepancy between an I.Q. test and academic achievement to determine a need for further support, the RTI model looks instead at how well a child responds to a tiered set of interventions. Other principles that are foundational to this model are collaboration between all the educational personnel, a willingness to try different strategies, and an increased involvement of parents.

Typically the model works on a three-tiered approach. In Tier One, students are in general education classrooms. Teachers assess students, adapt their curriculum, and use multiple strategies to promote learning. In Tier Two, students who need more support participate in supplemental services, small groups that occur during, before, or after school. Other school personnel may lead these. Finally, Tier Three is an intensive, individualized tutoring program for the student. The assessment process included screening, progress monitoring, and diagnostic reports.

This model increases the role of a classroom teacher in the identification and support of learners who may struggle. It is important that teachers understand underlying principles that will support struggling readers. Caldwell and Leslie (2009) identify seven important components of successful interventions for struggling readers as shown in table 6.3.

Table 6.3 Strategies for teaching special education learners

Strategy	Rationale
Emphasis on reading and writing for meaning	In the past, remedial reading instruction isolated skills through drills and worksheet. Struggling readers need encouragement and opportunities to make connections to their reading to understand that reading and writing are about meaning
Explicit instruction in the strategies that good readers and writers use	Strategy instruction makes explicit what goes on in readers' and writers' head. It demystifies literacy processes
Materials they can read successfully	Students should read texts that match their instructional levels and meet their interests
Consistency in lesson structure	Struggling readers benefit from anticipating a consistent structure so that their attention is focused on the content
Word study	Being able to accurately identify words is an important component in reading. When word study activities are integrated with comprehension activities, both are strengthened
A focus on fluency	Fluent readers are able to automatically sight identify words, decode unfamiliar words, and understand that the point of reading is comprehension. Fluency instruction helps struggling readers put these pieces together
Opportunities for small group instruction	Struggling readers need teachers who can listen to them and coach them as well as give them a chance to learn with and from peers

Differentiation for Purposeful Teaching

The common thread running through each of the preceding sections is the need for teachers to understand key characteristics and instructional supports for students, thoughtfully assess their students, and then adapt their instruction to the unique needs of the children in their classrooms. The key is effective differentiation of instruction, as Tomlinson (1999) discusses. According to Tomlinson, teachers can alter or differentiate the content, process, or product in their classrooms. The content is what she would like students to learn, the process is the activities planned to help students arrive at understanding or develop necessary skills, and the products are what student create to demonstrate and further develop their understanding of what they have learned.

Tomlinson (1999) identifies several principles that can guide teachers in thinking through and planning for differentiation in their classroom. Theses principles help teachers anticipate the diversity of needs and strengths of the learners in their classrooms. First, teachers need to focus on the essential concepts, strategies, and skills around the subject. Clarity about what is most important is key. Next, teachers need to carefully assess their students so that they can understand and appreciate their difference. An understanding that assessment and instruction are inseparable processes helps teachers carefully pinpoint and adapt instruction. In a differentiated classroom, teachers create authentic, meaningful work for all students, work that respects their abilities, provides opportunities for collaboration, and is able to take a flexible stance toward process and products in learning.

To plan for differentiation teachers can consider making changes, to meet the needs of their students, of the content, the process, and/or the product of learning. Tomlinson states, "Modify a curricular element only when 1. you see a student need and 2. you are convinced that the modification increases the likelihood that the learner will understand important ideas and use important skills more thoroughly as a result" (p. 11).

Building on Strengths

I was able to find some books to help improve his fluency, but at the same time I don't want the books to appear too easy for him because he also gets bored easily. If the books used for fluency are too easy, are on a topic Jesus enjoys and has previous information on, then he will most likely get bored. I would rather use books/information he doesn't know about for fluency and use the guided reading/read aloud books on topics that he would like to learn more about and topic we can expand on together.

Tara, Jesus' teacher, knows who he is as a reader. She is well aware of his need to become a more fluent reader, his present level of fluency, his interests, and how she is going to proceed with instruction. "Teachers must also be free to use material that allows them to connect what must be taught with what students can understand" (Darling-Hammond, 1997, p. 232). The freedom to plan a curriculum for her struggling reader, rather than having a curriculum guide determine the materials to be used and the lessons to implement, makes her time with the child more effective. Her assessments have informed her practice, one individualized to fit the strengths and the needs of the child.

What can other teachers learn from Tara? How are they grappling with finding and building on the strength of their students and being account-able for raising the test scores? Many of our teachers found a tension in planning for their struggling reader and writer; many were unable to trust their theoretical, pedagogical, and experiential understandings of literacy over prepackaged program instruction. An overemphasis on teaching to the portions of the reading curriculum that are testable on high-stakes tests has not only narrowed the curriculum but goes against the way teachers know is the best way to teach (Hollingworth, 2007), causing teachers to lose their confidence. We saw this when teachers would bring in worksheets, flash-cards, and components of their commercial reading programs into Literacy Space to use with their student, ignoring the room full of materials ready to be adapted to the strengths and needs of their child. Not given the flex-ibility at their schools, many students were uncomfortable planning, not just implementing, an instructional program for their children. Teachers need the opportunity to refocus and to reflect on what should inform their instructional practice.

> One struggle I have encountered working with children who struggle with reading is where to start and stop instruction. For example, when I was plan-ning my first three lessons for Kristen, I really struggled with figuring out exactly what to teach her. There are things that she knows and I want to build on her strengths but there is a lot of work that she needs.

Judy, as part of a Blackboard discussion, struggles with prioritizing instruction for Kristen but bases her decision making on the data she has collected about her tutee's literacy profile. She believes Kristen's needs as a reader must be at the center of the instructional decisions she will be making for her. The tutor knows Kristen's strengths and needs as a reader but questions where to begin instruction, how to prioritize the work that needs to be done. Without a commercial reading program with a scope and sequence chart neatly laying out a plan for instruction, teachers must use their assessment findings as a guide. By building on the child's strengths and interests, keeping them at the heart of her instructional plan, the teacher will be able to help Kristen develop the strategies she needs in order to per-form on grade level. However, the teacher needs to integrate assessment into her instructional plan to determine if she is meeting her instructional goals for the student.

Given time and flexibility, support and guidance, the teacher is making individualized instructional decisions for the child. Smith (2005), working with students in a reading assessment course, taught them to refrain from making judgments too quickly, to ask questions about the child's strengths and needs, to formulate hypotheses, and to gather assessment data logically. In doing so he found that teachers were able to restructure a naïve perspective of teaching (i.e., teaching is telling or teaching is fixing mistakes) into a perspective that puts inquiry at the heart of the teaching and learning process. Yet Lenski and Nierstheimer (2006) found that a field experience and acculturation into an unsuccessful school program can override the often valid and important concepts learning during course work. The tension between the two needs to be mediated by the teacher educator. We found that, like Smith, our teachers needed time to make instructional decisions, space to be reflective in their practice, and to keep the child at the heart of their decisions.

Teaching for the Long Term

Teachers and children thrive when they are given an opportunity to make real choices. Ivey (1999) found that students want to be and can become good readers when they are motivated with real purposes for reading, provided with a wide variety of materials of their interests, and the opportunity to collaborate with their peers. Moving teachers away from a deficit model and being able to trust their assessments as a guide to planning instruction is a shift fraught with tension and anxiety. Yet, once accomplished, teachers are able to build on strengths. Kristen's teacher, learning to trust her interpretation of the data the assessments yielded, coupled with her familiarity of the child, said, "I found myself pouring over my professional books and finding many good ideas that I want to bring to my work with Kristen."

> Often times when working with Juan in the past and during one session I had with him this semester, I find that there are so many things I can work on and in the end I feel that I really didn't do anything at all. I need to work on quality as opposed to quantity.

The instructional direction we take each day influences the beliefs children have about reading. For many, the instructional direction becomes a fork in the road and both paths may not clearly lead us to where we want to go. Verbalizing the struggle, one of our teachers said:

> One of the things I struggle the most with is figuring out what would be the most important or memorable teaching point to address during my lesson with a student. There are times when I have worked with students who display lack of skills in many areas and making this decision becomes very challenging. I want to be able to draw on their attention to something that will be effective for them to use in their future reading experiences with me as well as in their classroom.

This blackboard posting clearly illustrates a dilemma for many teachers. What do I do in the classroom that will make a difference for my students down the road? Teaching for the long term can mean putting aside the quick and easy stuff to teach, a Band-Aid approach, or teaching to deficits. Teaching for the long term means teaching to strengths and also one that puts inquiry at the heart of the teaching and learning process (Smith, 2005). Rather than looking to fix a reading problem, putting the onus on the student, teachers need to make adaptations in their instruction to meet students' literacy needs (Vogt & Shearer, 2007).

Revisiting the Field

Drawing Out Eduardo

Carla wanted to trade tutoring assignments when her initial meeting with Eduardo was baffling and difficult for her. Instead, I offered to sit with her and Eduardo and support her for the rest of the session. The table was filled with magnetic letters, cards, books, and crayons. I suggested that we move all the materials away so that we could have some space, and invited Eduardo to help me. After we put the letters away, we returned to *The Hungry Caterpillar* book. I asked Eduardo if he wanted to read the book again and he agreed. This time, I advised Carla to offer Eduardo a chance to talk about the pictures and foods, as the caterpillar eats. I asked Eduard to tell me if he liked the foods the caterpillar was eating—thumbs up or thumbs down. Sitting on the edge of his seat, Eduardo participated eagerly. Thumbs up for pie! Thumbs up for strawberries!

After reading we asked Eduardo to draw about the book. Carla and I both were amazed by the details as Eduardo drew the caterpillar munching through fruit. Carla whispered to me, "He so good at drawing. I wouldn't have guessed that." Feeling more connected now to Eduardo, Carla continued with her assessments. She was more willing to suspend her Special Ed. label as she began the work of finding out who he was as a reader and writer.

A Partner for Oscar

Rosa struggled with figuring out the best way to challenge Oscar with appropriate reading work but did not want to further separate him from his peers. As part of a getting-to-know-you activity, she surveyed her students about their interests—from favorite movies to foods to books. Reviewing the surveys she discovered that Christian, a popular but quiet student, put down Harry Potter as his favorite movie, although as a reader on Level G he was far from being able to independently tackle the book. Rosa decided to ask Oscar and Christian to be reading buddies. She hoped that their mutual interest in Harry Potter would help them forge a friendship (something Oscar needed) and provide an opportunity for them to work together on a Harry Potter project. She allowed them time to work on an independent project, to teach the class about Harry Potter. Oscar and Christian created a Harry Potter PowerPoint to share their information with the class. Each boy took turns telling about the characters, author, and differences between the books and movies. When the presentation ended, to wild applause, Rosa believed she was on the way to having a good year with Oscar.

Balancing Test Prep and Fun

Janet struggled in planning instruction for Karina. She tried to work on test prep activities the school wanted her to focus on and found Karina lost interest and became bored. Using test prep materials and focusing on Karina's future performance on upcoming formal assessments interfered with Janet getting to know Karina as a reader and as a person. It became difficult to assess Karina's strengths and vulnerabilities as a reader. Janet refocused her attention on helping Karina develop as a reader. She used the data she had available with her to guide the instructional decisions she made. She let Karina's real needs inform her instructional plan. From an interest interview and informal conversations, she had learned that Karina was an avid dancer, who performed in a Mexican dance troupe on the weekend. Janet searched for articles about dance and biographies of dancers to incorporate into her instruction. By working on Karina's basic needs and matching activities to her needs, Janet made real progress. By teaching for the long term, Karina had the opportunity to become a more strategic reader, rather than being taught to pass a test.

Strategies and Tools

Moving beyond the Label by Knowing and Teaching All our Students

Classroom teachers sign on to teach all of the children in their classes, regardless of their educational labels. Good teachers strive to get to know their students and understand, in a deep and informed way, educational labels they may have; accept that it takes time and effort to meet the diverse needs in their classroom but that it is critical for children's success to do so; are willing to extend their own professional development by being researchers themselves to strengthen their own learning; and acknowledge and work through their own worries and self-doubts about their teaching so that they can focus on the children in their classrooms.

Acknowledge and Move beyond the Label

Carla and Rosa felt uncomfortable and challenged by children whose literacy performances were unfamiliar. Janet felt pressured to conform her instruction to a single test. Despite our professional training and desire to teach all the children in our classes, there will be times when we feel unprepared or out of our depth. This is a time not to give up but to work through your concerns, to work through discomfort. By acknowledging your emotional and professional self-doubt about their performance you will be able to move beyond panicking or avoiding really addressing your students' needs. One important way to begin this process is by recording and then examining the language you use to describe the children who challenge you. In the Tool for Practice *Words and Labels* you can use the chart to list words that come to your mind when you think of specific children and then try to unpack why you might have used that word, the implications of the word choice, and what it tells you about where your instruction needs to move.

Know Your Students

It is widely recognized that teachers need to take time to have conversations with their students. Many teachers have writing and reading conferences already in place. Yet many novice teachers do not know that the conference time is more than just a time for teaching students individually. They are not fully aware that they can use this time to learn from the children until teachers record their conferences with students and analyze their own conference transcript. "It's amazing," wrote Fatima, a second grade teacher, "I noticed that I kept on asking questions and I didn't even listen to what the child was trying to tell me." Reflecting further on her conference, she realized that her student's story was clearly underdeveloped. Instead of helping the child to develop the content more, she noticed that she seemed to be fixated on getting the child to use the five senses. Retracing her steps allowed her to see that she needs to truly listen to the child's story so that she could teach to the child's intentions.

For the face-to-face teaching to be effective, teachers must learn from children shoulder-to-shoulder. Teachers should expect to learn from the child every time they pull up a chair to sit beside her. Ask questions and really listen for answers, like a researcher would. Get on a line of inquiry. What does this child know? What is she good at? Why does she not comprehend the text while her running record shows an accuracy rate of 95 percent? What can he teach me that I can teach the whole class? Only when we expect to learn from our students, only when we are in the habit of learning alongside our students, can our teaching, our implementation of best practice, methodologies, and materials be effective in teaching the same children who are also our teachers.

Getting to know the children in your classrooms is key toward being able to move beyond labels. This is particularly important when you feel puzzled by a student or have a student with an educational label or IEP. Carla was better prepared to work with Eduardo when she took the time to learn about not only his academic needs but also his unique interests and abilities. The Tool for Practice *Student Profile* can help you record information about your students. By taking the time to find out about your students' interests, strengths, views of reading and writing, learning style, you will gain important insights into who they are and find places to launch your instruction. Teachers who tutor in our Literacy Space practicum, like Carla, find that the process of finding out about the child they are working with helps them to understand the children. Classroom teachers like Rosa, find that using the *Student Profile* helps them in differentiating instruction and forming flexible groups.

In addition to getting to know your students' profiles, to better move beyond labels and differentiate instruction it is important to know as much as possible about their specific literacy strengths and needs. The Tool for Practice *Diagnostic Profile* can help you track your assessments of the children.

Children's interests can be more important than the readability of a text. It is often difficult for teachers to be responsive to students' interests and needs and most critically, invite students to collaborate in planning their instruction. Yet when Kristy, a resource room teacher, recognized third grader Al's interest in the New York Giants football team, her sessions with him broadened. She brought in a newspaper with information about obtaining New York Giants stickers. Al enthusiastically read the advertisement, which directed readers to a website with additional information about purchasing an album. He asked to do more research. Although not in Kristy's plan, she seized on his interest and they moved to the computer where Al read avidly.

As teachers move to a more responsive, collaborative stance, the children, too, become more engaged active participants in their own learning. Children state they like to read when they are provided materials with an array of interests and levels of difficulty; they are motivated by real purposes of reading to become good readers (Ivey, 1999). One of the teachers we work with frequently expressed concern about her student's lack of engagement and motivation. After using a picture book with accompanying materials, she stated, "She [the student] had a great time with this story as we used the props as we read. The props helped her read the names of the animals."

Reach Out and Research

It is almost certain that you will encounter children with unique learning needs throughout your teaching career. Taking an active, inquirer stance will help you to learn more about those learners. The Tool for Practice *Learn More!* is designed to help you identify resources to support your work with diverse learners. Resources can include experienced teachers, related professionals such as school speech teachers or ESL teachers, professional texts, professional organizations, and websites.

Ellen wrote "Teaching reading to non-English speaking children was a huge challenge for me. I felt like I had no starting point. I got help from the ELL teachers. We exposed the children to books in their native language at the start to give them confidence. I labeled everything in my room and used pictures on the word wall." By asking the ELL teachers for help, Ellen initiated a productive professional collaboration.

Effective Literacy Teaching

We strongly believe that effective literacy teaching demands a deep understanding of literacy development and instructional practices coupled with a deep and continually growing understanding of the students. Blair et al. (2007), in their review of literature about reading teachers, synthesized five common features:

1. Assessing students' strengths and weaknesses.
2. Incorporating some explicit teaching.
3. Providing students with opportunities to learn and try out developing abilities in authentic reading activates.
4. Ensuring children are engaged in reading tasks.
5. Believing in one's teaching abilities and expecting children to be successful. (P. 433)

Tool for Practice: *Words and Labels*

In this chapter and in chapter two, we've asked you to think about the words you use to talk about children. You can use the tape recording of your lesson or look at your report cards or rubrics. This tool (table 6.4) gives you a chance to analyze your language.

Table 6.4   Examining words and labels

Adjective you use	Child referenced in the adjective	Potential label	Educational issues

Table 6.5   Student profile

Interests outside of school
Views of reading/writing
Academic strengths
Academic needs
Family feedback and concerns

Tool for Practice: *Student Profile*

Getting to know your students is an important process that begins when they enter your classroom. This tool (table 6.5) is a summary sheet where you can record the information discovered through reading interviews, interest inventories, anecdotal observations or "kid-watching," and conversations with families. Create a sheet for each student in your class. You can add comments and observations throughout the year.

Tool For Practice: *Worksheet for Diagnostic Profile*

This tool (table 6.6) provides a chart for you to record your findings as you assess your students' reading abilities. By looking at the variety of literacy areas and by identifying students' strengths and areas of concern, you can get a fuller picture of your student's reading behaviors in order to plan instruction.

Child's Name_____ Grade_____

Date_____

Table 6.6   Worksheet for diagnostic profile

Literacy area	Tools used	Rationale for tools	Strengths	Areas of concern
Oral language				
Concepts of print				
Phonological awareness				
Decoding				
Sight word knowledge				
Fluency				
Comprehension				
Strategic knowledge				
Writing				
Spelling				

What organizations would offer information?	Who are the teachers who can help me?
_____ _____ _____ _____ _____ _____ _____	_____ _____ _____ _____ _____ _____ _____

Who in my building has expertise?	What books can I consult?
_____ _____ _____ _____ _____ _____ _____	_____ _____ _____ _____ _____ _____ _____

Figure 6.1   Learning more about my students.

Tool for Practice: *Learning More!*

As a new teacher it can feel overwhelming to identify and understand the needs of diverse learners in your classrooms. While your initial teacher education program may have provided an introduction to the strengths and needs of diverse learners, as well as some practical, instructional approaches, you may feel you need further education.

This tool (figure 6.1) is designed to help you identify different sources of information so that you can take charge and continue your professional development and learning.

Suggested Resources

Caldwell, J.S., & Leslie, L. (2009). *Intervention strategies to follow informal reading assessment: So what do I do now?* (2nd ed.). New York: Allyn and Bacon.

Dillon, D.R. (2000). *Kids insight: Reconsidering how to meet the literacy needs of all students.* Newark, DE: International Reading Association.

Dudley-Marling, C., & Paugh, P. (2004). *A classroom teacher's guide to struggling readers.* Portsmouth, NH: Heinemann.

Duffy, G.G. (2003). *Explaining reading: A resource for teaching concepts, skills, and strategies.* New York: Guildford.

Freeman, D., & Freeman, Y. (2007). *English language learners: The essential guide.* New York: Scholastic.

Halsted, J.W. (2002). *Some of my best friends are books: Guiding gifted readers from preschool to high school.* Scottsdale, AZ: Great Potential Press.

International Reading Association. www.reading.org.

Tomlinson, C.A. (1999). *The differentiated classroom: Responding to the needs of all learners.* Alexandria, VA: ASCD.

References

Allington, R.L. (2006). *What really matters for struggling readers: Designing research-based programs* (2nd ed.). New York: Pearson, Allyn and Bacon.

Blair, T.R., Rupley, W.H., & Nichols, W.D. (2007). The effective teacher of reading: Considering the "what" and "how" of instruction. *The Reading Teacher, 60,* 432–438.

Caldwell, J.S., & Leslie, L. (2009). *Intervention strategies to follow informal reading assessment: So what do I do now?* (2nd ed.). New York: Allyn and Bacon.

Darling-Hammond, L. (1997). *The right to learn: A blueprint for creating schools that work.* San Francisco, CA: Jossey-Bass.

Dudley-Marling, C., & Rhodes, L.K. (1996). *Readers and writers with a difference: A holistic approach to teaching struggling readers and writers.* Portsmouth, NH: Heinemann.

Freeman, D., & Freeman, Y. (2007). *English language learners: The essential guide.* New York: Scholastic.

Guag, M. (1984). Reading acceleration and enrichment in the elementary grades. *The Reading Teacher, 37*(4), 372–376.

Hollingworth, L. (2007). Five ways to prepare for standardized tests without sacrificing best practice. *The Reading Teacher, 61,* 339–342.

Ivey, G. (1999). Reflections on struggling middle school readers. *Journal of Adolescent and Adult Literacy, 42,* 372–381.

Krashen, S. (2003). *Explorations in language acquisition and use.* Portsmouth, NH: Heinemann.

Lenski, S.D., & Nierstheimer, S.L. (2006). Teacher preparation programs offer targeted field experiences in literacy. In S.D. Lenski, D.L. Grisham, & L.S. Wold (eds), *Literacy Teacher Preparation: Ten Truths Teacher Educators Need to Know* (pp. 44–53). Newark, DE: International Reading Association.

Lindholm-Leary, K. (2000). *Biliteracy for a global society: An idea book on dual language education.* Washington, DC: National Clearinghouse for Bilingual Education.

Moon, T.R., Callahan, C.M., & Tomlinson, C.A. (1999). The effects of mentoring relationships on preservice teachers' attitudes toward academically diverse students. *Gifted Child Quarterly, 43*(2), 56–62.

Moore, M. (2005). Meeting the educational needs of young gifted readers in the regular classroom. *Gifted Child Today, 28*(4), 41–47.

Moss, M., & Puma, M. (1995). *Prospects: The congressionally mandated study of educational growth and opportunity.* Cambridge, MA: Abt Associates.

President's Commission on Excellence in Special Education (2002). Washington, D.C.

Renzulli, J.S. (1998). A rising tide lifts all ships: Developing the gifts and talents of all students. *Phi Delta Kappan, 80*(2), 104–111.

Sleeter, C.E. (2001). Preparing teachers for culturally diverse schools. *Journal of Teacher Education, 52*(2), 94–106.

Spear-Swerling, L., & Sternberg, R. (1996). *Off track: When poor readers become "learning disabled."* New York: Westview.

Tomlinson, C.A. (1999). *The differentiated classroom: Responding to the needs of all learners.* Alexandria, VA: ASCD.

Tomlinson, C.A. et al. (1994). Practices of preservice teachers related to gifted and other academically diverse learners. *Gifted Child Quarterly, 38,* 106–114.

Townsend, J. (1996). Big books: Links to literacy for everyone. *Perspectives in Education and Deafness, 14*(3), 23–24.

Vogt, M.J., & Shearer, B.A. (2007). *Reading specialists and literacy coaches in the real world.* New York, NY: Allyn and Bacon.

Vosslamber, A. (2002). Gifted readers. Who are they and how can they be served in the classroom? *Gifted Child Today, 25*(2), 14–21.

7

Teaching Effectively Means Learning from Our Students

I guess what I'm trying to say is that I don't know how much I really should be sticking to my plans. The teacher in me says, "be organized, plan and follow the plans or you'll miss something important." But my heart tells me to follow my instincts and do what I feel works at the moment. Does anyone else struggle with this?

—(Jennifer, a third grade teacher)

Jennifer's dilemma is one that novice teachers face every day. The teacher in her tells her to follow a well-organized plan. She had designed the plan carefully around the district's curriculum goals. She had included the content she knew the children would need to move on to later topics. But she cannot ignore the subtle signals the students are sending her: they seem to be bored and disaffected, many keep glancing at the clock on the wall, and twice they needed to be reminded what they were supposed to do. Should Jennifer implement the plan she designed to meet the instructional goals she had previously determined her students needed? Or should she abandon the plan and listen to what the children are telling her they need, instructionally, at the moment?

Jennifer's plan defines a comfort zone, where the objectives are clear and the path to the objectives predictable. Yet she knows that she must teach the class, which means being open to negotiating the path with them. Such a path could lead to uncharted territory, to the edge of her comfort zone.

Challenge and Opportunity

A tension exists for teachers when the planned lesson, either scripted or self-designed, and the implementation of the lesson does not seem effective for student learning. Sometimes the scripted lesson or prescribed methodology is not appropriate for the particular group of students for which the lesson was designed. Students may not have the appropriate background knowledge, requisite skills, or interest in the topic as it is presented. Other times teachers do not know how to embrace the teachable moment and move away from what was planned when students indicate lack of understanding. Teachers have not planned nor do they know how to accommodate the lesson midstream. Occasionally, teachers do not know their students well enough, do not know their interests, capabilities, and background enough to incorporate or accommodate it within a lesson.

In many schools teachers are required to follow scripted lessons. When they see their students falling behind they begin to question the effectiveness of the script they have been given. They also question what they believe to be best practice when it is not incorporated as part of the script or methodology they have been told to follow. The frustration teachers feel and the tension created in trying to follow a plan is real. They often are insecure in adapting or supplementing the material they have been given and the plan they are required to follow. They feel anxious in diverting from the plan fearful of leaving important concepts and strategies out.

For teachers and literacy coaches who are starting out, finding themselves in a gap between the known and unknown can produce tension and discomfort. Too often they find themselves falling back on what they know and what is comfortable rather than moving ahead and exploring what might be best for the student. For example, when students do not appear to comprehend a reading, teachers will rely on questions to which the answers are explicitly stated in the text thinking the students are understanding the material when they answer correctly yet fail to ask questions that are more thoughtful and require a deeper understanding on the students' part. They ask questions to which they already know the answer. All teachers carry with them a select number of strategies with which they are comfortable and implement them in a variety of settings for a variety of instructional purposes. They find the strategies comfortable, reliable, and readymade.

As discussed in chapter six, teachers need to know their students in order to teach them and many teachers are not comfortable allowing their children to inform their practice. As thoughtful practitioners, teachers need to make decisions in adapting and adopting curriculum to meet the needs of their students.

How can teachers use information about individual children to inform whole class instruction? How can teachers adapt, supplement, or rewrite lesson plans to accommodate the group of students in a class? Negotiating the curriculum, not disregarding it but keeping it within reach, and being able to do what is best for the children often causes tension for teachers. Fear about meeting standards, keeping to a script, seeing the alternatives to implement best practices often require shifts in thinking. It means being able to abandon a plan when students' capabilities indicate they are not ready or equipped to move to the rhythm of the planned lesson. It means recognizing when the lesson is failing, to seize that moment and turn the lesson around. Within the classroom routines, teachers need to learn from their students in order to teach them, to honor the teachable moment within a prescribed or self-designed plan.

Stories from the Field

Maria was in her second year of teaching fourth grade in an overcrowded school. The majority of children came from single parent homes where the primary language was English, were eligible for free or reduced lunch, and were below grade level as indicated by their last standardized tests. As a second-year teacher Maria felt grateful to have the structure of a reading series. It provided her with the pre-reading, during reading, and after reading suggestions she needed for planning her lessons. She kept pace with the series thinking this would ensure her students' reading success. But her students could not keep up with the pace set by the program and seemed disengaged from the readings. It was only November and Maria felt she, and her students, were falling behind.

The next story in their textbook was a nonfiction piece about cave paintings. She reviewed the pre-reading questions, read over the story and the questions that followed the text. The extension activities looked great but she knew she would not have time for them. It was hard enough getting them to read the passages and answer the questions at the end.

Maria began the lesson by telling the children that the new reading story was nonfiction and asked the children what that meant. She was pleased when they told her that it was true. She continued the pre-reading questions by asking what they knew about caves and if anyone had ever been to a cave. Again, she was pleased when one or two children told her that bats lived in caves and they had learned about caves in science. She asked them to read the story. But after reading when she asked them about the paintings discovered in caves, the class was silent, clearly confused. Maria was at a loss. She had followed the script in the text and asked the pre-reading questions to which some children had provided answers. What happened?

In a first grade classroom, Linda gathered her Level G readers at the back table for a guided reading lesson. The book was about a visit to the optometrist. She asked the four children in the group what an optometrist was from looking at the cover of the book. She reminded them that as a class they had recently been to the nurse for something. "What did we do when we went to the nurse last week?" she asked the children. It was clear that she wanted them to recall having their eyes tested and how they were asked to read from a chart. But students told her that they stood in line, they had to be quiet, and other responses that had nothing to do with having their eyesight tested. Yet Linda kept asking the same thing over and over again, clearly frustrated and frustrating the children until they finally remembered having their eyes examined.

Before reading, she elicited from the children strategies that they could employ when they came to an unknown word. The children began to read silently as she moved from child to child, had each read aloud, and recorded notes on their reading. Almost all students, none of whom wore corrective eye wear, got stuck on the work "frames" and using the picture came up with the word "glasses." Linda became more and more frustrated with her students. The strategies they were employing were not working, they were not comprehending the story, and Linda, having followed the guidelines given to her for a guided reading lesson, was at a loss.

Maria and Linda found themselves at the edge of their comfort zone. Both were baffled as to why their well-planned lessons just didn't work they way they had expected.

Research and Reflection

Based on many research studies on effective classrooms, Cunningham and Allington (2007) found some common characteristics of classrooms and their teachers that provided the best instruction in reading and writing for children. Not one of the exemplary teachers used only one program or set of materials; instead of using worksheets and repetitive drills, children were doing a lot of reading and writing, and the teachers took the children's interests and needs into account. Teachers devoted time to model, demonstrate, emphasize meaning and integrate reading and writing into other content areas. Knowing these characteristics and adopting them to put into practice is difficult for many teachers. They do not feel they have the freedom or support needed to implement these effective practices, indeed best practices, in their own classrooms.

Many teachers are handed a curriculum, some scripts, texts, and methodologies mandated by the school or district. Others struggle when the plans they have designed, consistent with their school's curriculum and objectives fail their students during implementation. What teachers believe to be most effective sometimes gets lost. Best practices may fade when contradicted by their experiences in classrooms and the way in which schools implement literacy instruction (Toll et al., 2004).

The Entry Point: Finding the Overlap between the Curriculum and Students' Needs

For some teachers, the curriculum itself presents a problem. Like Maria, most teachers have been taught that effective teaching follows a linear path: planning, implementing, assessing. Usually assessing meant assessing how effective the teacher was in implementing the lesson. Teachers were all taught to focus on their teaching, which they carefully demonstrated on their lesson plans. The plans read like recipes of what was to be taught, how it would be taught, the materials that would be needed, the objectives to be met, and a step-by-step process of what would be said, and what the children would be asked to do. Lesson plans left little room to really pay attention to how the students were learning. Commercial packages have left teachers with a false sense of security allowing them to think that if they covered all the lessons they were good teachers (Allington & Cunningham, 2007). Formative assessment is found at the end of the lesson and not built in along the way. It then lacks information needed to move students' learning forward.

Teachers, reading specialists, and school administrators need conversations that help them interpret the curriculum guidelines together allowing time to establish and implement curriculum changes together to meet diverse needs of children during whole class instruction (Darling-Hammond, 1997; Fullan, 2003). Teachers' instructional decisions are often affected by the school guidelines leading them to implement practices they believe are not best for their students (Davis et al., 1993). Altwerger and her colleagues (2004) found teachers "are less than open to exploring strategies and concepts that they don't believe they can use in their classrooms and are apt to question the credibility of strategies that are not part of a packaged program" (p. 126). And Anderson (2009) found that elementary school teachers relied on generic teacher-centered practices rather than on student-centered ones.

Teachers need the opportunity to make connections that allow for the creation of a series of experiences for their students' needs and those imposed by outside standards. "To be effective, teachers must meet students *where they are,* not where an idealized curriculum guide imagines they should be" (Darling-Hammond, 1997, p. 232). Similarly, Lipson and Wixson (1997) posited that one-size-fits-all programs for instruction are antithetical to effective practice. The unique experiences of a group of children, the strengths and vulnerabilities that they bring to the learning situation, can provide information about how the children in a classroom acquire literacy. Luke (1994) suggested the teacher and reading specialist ask how information informs best instruction for that particular child or a group of children. Teachers need to be responsive not only to the interests, strengths, and vulnerabilities of each individual child, but to a class of children, and not with a preprogrammed response or by using an old reliable method, but by creating a unique series of experiences for the children in order to move the children's literacy development forward. It requires a thoughtful look at student work and the ability to listen to student talk carefully. One of the characteristics of effective teachers was they would regularly assess how the children were progressing and adjusted their instruction based on those assessments (Cunningham & Allington, 2007). In planned-for instruction, as termed by Heritage (2007), teachers decide while planning instruction how and where they will elicit student understanding. By asking students to explore ideas and respond during instruction, the teacher can gather valuable assessment information about student learning. Darling-Hammond (2007) stated, "A skillful teacher figures out what students know and believe about a topic and how learners are likely to 'hook into' new ideas" (p. 98).

Teachers often ask themselves, "But how do I know what my class needs? Doesn't the curriculum provide the structure and sequence I should follow to help my students develop the strategies they need and meet the standards given to me?" According to Dole (2004), to ensure students are achieving the teacher must use effective and efficient informal assessments to monitor progress of each child. By employing periodic interest inventories, discussing concepts that will be presented in the literacy curriculum, and monitoring the independent use of instructional strategies, patterns of interests, needs, and strengths can be determined and lessons can be focused on the needs of the children in the classroom.

Dearman and Alber (2005) suggest the establishment of study teams, which provide teachers with the opportunity to improve their practice by sharing their reflections, knowledge, and solutions. Teachers need to seek out professional partnerships with colleagues where schools have not given them the support or time they need to meet. Negotiating the dichotomy of curriculum standards and best practice in study teams promotes understanding of the curriculum demands without having to sacrifice beliefs in instructional practice that meets the needs of the children. According to Bean and her colleagues (2002), a solid understanding of the classroom curriculum needs to be in place in order for teachers to be able to modify instruction to meet the needs of students and meet the required instructional goals. Teachers must have a working knowledge of the curriculum in order to connect their students with sources of information to allow them to explore ideas, acquire information, and to problem solve (Darling-Hammond, 2007).

Rather than quickly finding a solution to the dilemma of conflicts and tensions existing between what is mandated and what is believed as best and responsive practice, Toll and her colleagues (2004) suggest making the conflicts visible, in order to create spaces in instruction for complexity and competing desires. They go on to urge teachers to explore the power available to them and to look to both the mandates and best practice to discover any overlap.

Learning about Students before Teaching

Children's interests can be more important than the readability of a text. It is often difficult for teachers to be responsive to students' interests and needs and most critically, invite students to collaborate in planning their instruction. As teachers move to a more responsive, collaborative stance, the children, too, become more engaged active participants in their own learning. Children state they like to read when they are provided materials with an array of interests and levels of difficulty and they are motivated by real purposes of reading to become good readers (Ivey, 1999).

It is widely recognized that teachers need to take time to have conversations with their students. In those conversations teachers need to actively listen and ask questions of their students. As Routman (2000) stated, "When we question students, it is not so much to test their knowledge but to find out what they understand, are curious about, want to know, need to know" (p. 466).

Many teachers have writing and reading conferences already in place. Yet many novice teachers do not know that the conference time is more than just a time for teaching students individually. They are not fully aware that they can use this time to learn from the children and use that new knowledge in preparing whole class instruction. As discussed in a previous chapter, teachers can record their conferences with students and analyze their own conference transcript. By looking for patterns across conferences, the teacher is using knowledge about the students to inform her practice. She can determine what, as a class, they need and then adjust her practice to meet those needs. The comments a teacher jots on students' papers can also be used for planning whole class instruction. Too often the papers are returned to students and valuable instructional planning information is lost (Allington & Cunningham, 2007). Recording the notes made on student papers and analyzing them for patterns can be used to inform instruction. It is about asking the right questions of the teachers' comments and notes about children.

What can the child teach me that I can teach the whole class? Only when we expect to learn from our students, only when we are in the habit of learning alongside our students, can our teaching, our implementation of best practice, methodologies, and materials be effective in teaching the same children who are also our teachers.

Learning to Teach Our Students

Long before teachers learn about teaching in their formal course work, they have already had experiences with their families and schools and have a set of perceived and acquired ideas about teaching (Hollingsworth, 1989; Lortie, 1975). In a case study by Theriot and Tice (2009) they reported that teachers often fell back on instructional decisions incompatible with their stated beliefs about literacy when they ran into obstacles. They stated, "A deep understanding of the philosophy and theory that undergrid an approach, however, can make it possible to arrive at instructional decisions that are consistent with an approach as a teacher deals with an unknown" (p. 71).

Teachers tend to be loyal to their instructional strategies but often do not realize that all are not equally effective. As Greenwood and Maheady (2001) suggested, teachers must ask themselves to identify the evidence that a particular approach has successfully improved student learning. A tension exists when teachers feel the need to improvise during a lesson as they become tuned into their students and discover the students are not learning from the planned lesson. "If teachers investigate the effects of their teaching on students' learning and if they read about what others have learned, they become sensitive to variation and more aware of what works for what purposes in what situation" (Darling-Hammond, 2007, p. 95). Teachers must have a wide variety of clear and effective teaching practices from which to draw upon during these circumstances. Good teachers have a large repertoire of strategies from which to draw on when entering an instructional situation in order to personalize instruction (McGill-Franzen & Allington, 2005; Roller, 1996). Teachers must be able to use a variety of teaching strategies to accomplish different goals and be able to assess their effectiveness on student learning.

Teachers are supplied with curriculum guides provided by their state and district. They are also provided with teacher manuals that accompany textbooks used in the various grade levels. Each outline curriculum goals and objectives the teachers need to meet, strategies children need to develop, and content children need to own by the end of a grade. How teachers use these to plan for their group of students should depend on the abilities, needs, and background of the students. Children will always vary in their ability levels, class-by-class, year-by-year. And because of this, teachers are continuously involved in making innumerable, practical decisions about instruction daily. The curriculum guides do not provide clear rules that can be listed in the manual or applied in a systematic way from one learning situation to the next. Teachers know that learning is not a one-way process but involves watching the children for student learning. "It is the application of accumulated skill, wisdom, and expertise in the specific and variable circumstances of the classroom which defines much of the teachers' professionalism" (Fullan & Hargreaves, 1996, p. 19).

Effective teaching is a recursive process of learning, teaching, learning, teaching. The "learning" consists of learning from our students, to understand what they bring to our carefully designed lessons and what they need from us, namely, information that helps them learn effectively. The "teaching" consists of finding the overlaps between the curriculum and the students' needs and prior understandings, and of addressing children's needs through both explicit instruction and guided practice. They should be reciprocal activities and an integral pattern of teaching and learning. It means stepping to the edge of the comfort zone of carefully designed plans and allowing the students to inform practice. It is about the opportunities we afford the student. "Learning depends on the active engagement of the learner. It is what the learner does that is learned, not what the teacher does" (Anderson, 2009, p. 416).

Teachers need to be encouraged to shift away from a narrow range of instructional materials and strategies that attempt to patch or fix areas identified as weaknesses. Teachers need the resources and support to develop lesson plans that incorporate specific and explicit instruction through their careful analysis of challenges presented in authentic literature. "Teachers must be free to use material that allows them to connect what must be taught with what students can understand" (Darling-Hammond, 1997, p. 232).

Revisiting the Field

Maria read the next lesson in the reading series. It was about the beginning of the Wells Fargo Bank and the California Gold Rush. Maria knew her east coast, urban students had little prior knowledge to prepare them for this text and, without careful planning, little interest. The pre-reading questions provided in the text would not tell her if the children had the appropriate background for the story. Maria needed to find an entry point to develop a real reason for reading, understanding, and appreciating the story.

Maria planned her lesson and began with a discussion about money and banking. She used current newspaper articles to pique the students' interests about present day banking. She then moved the students' interest to an historical account of banking.

Using the map of the United States and knowledge about the geography of northern California from her previous Social Studies lessons, Maria was able to relate the students' prior knowledge to new knowledge presented in the reading lesson. By creating a pre-reading activity to which her students could relate and by building the background the students needed to understand the story, Maria was able to move the children's literacy development forward. This kind of planning enabled Maria to listen to what her students needed to learn alongside of them in order to teach them. She was also able to keep to the curriculum mandated by the school. She was able to adopt a perspective that allowed for a wider range of literacy practice within the confines of the curriculum.

Linda tried to implement the guidelines she had been given for guided reading lessons. She knew the goals and objectives she needed to meet and had been planning with those in mind. Linda had not been considering what her students brought to any given reading in terms of their background knowledge or interest. She was also not focusing instruction on a specific reading strategy or determining which strategies the children already owned. This made her guided reading lessons frustrating for both the children and for her. She was focused on what she should do and not what the children should do.

After conferencing with a colleague, Linda learned to plan with her students in mind and put her focus on what they should do during the lesson. She thought of alternate methods of getting to the information she wanted them to own by changing her own methodology. She explored a variety of formats for a guided reading lesson before planning. In thinking back to her lesson on the optometrist she discovered that a shared reading of the book to begin the lesson would have allowed the children to learn the special vocabulary of the text. Stopping at predetermined times during the reading to discuss the concepts would have provided additional opportunity for comprehension. Linda now builds in assessment opportunities within each lesson so she knows at predictable increments if her students are "with her" and, if not, embracing the teachable moment with what they seem to need in order to move the lesson forward.

Both Maria and Linda found themselves at the edge of their comfort zone when their planned lessons fell flat. The lessons, one scripted and one planned around guidelines, were not meeting the needs of the children. This surprised them both since they had so carefully prepared to implement their plans and were sure of the plan's effectiveness. When the signals from the students indicated they were "not getting it" neither Maria nor Linda knew what to do.

They needed to be reminded that prior knowledge is important to comprehension and discussion and to consider the prior knowledge their students brought to the lesson. They learned that teaching effectiveness wasn't measured by what they had planned but what the students were learning. And when it was clear that students were not involved with the lesson in the way they had planned, they knew that they needed new instructional strategies. They had to allow the children to inform their practice.

Strategies and Tools

Think of the process of teaching and learning as a triangular figure (figure 7.1). The novice teacher tends to focus on herself and the content, and less attention is paid to the students, an often neglected part of the triangle. Through most of her undergraduate experience, the focus was on what she planned, what she implemented, her classroom management, and lesson implementation. Much classroom instruction was given to the design of unit and lesson plans. The very nature of their design has to do with content and what the teacher will do, the materials she needs to implement the lesson, the questions she will ask, the activities she will implement and very little focus is on the assessment of student learning. During student teaching, her success was based on an observation of how well she performed in the classroom. It was more about classroom management than about student learning.

Assessing teacher effectiveness should be more about student learning. More attention needs to be on the students in the triangle. As the student teacher moves into the teaching role, her previous experience needs to be rebalanced to include more attention to student learning. The goal of this section is to give teachers early in their careers strategies to revise and fine-tune their own teaching in response to student learning. Our intention is to help the teacher, early in her career, take common strategies and make them concrete, visible methods to access student learning, and to help transition the teacher into making those strategies readily available to her while engaged in teaching and learning.

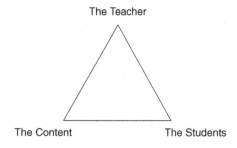

Figure 7.1   The teaching-learning process.

Monique, a fifth grade teacher, already knows the value of learning from her students. She said,

> We have guidelines we are required to follow and usually very few opportunities to be creative. Our students are at such different levels it might seem impossible to plan games, inquiries and still get to everything we are expected to do. We learn from our students but we usually are so flustered we rarely take the time to embrace what they've taught us and to recognize from it...Many times we don't have that luxury in our classroom and we lose many "teachable moments" because of it.

Observing for Student Learning

The triangle given in figure 7.1 reminds us of the challenge to pay attention to all three angles at once: the content, the teacher, and the students. We need to learn to watch for the student angle. But the power of observation develops over time. As mentioned previously, teachers make innumerable decisions throughout the day and most based on observation of the students in the classroom. There are strategies that teachers who are starting out can employ to help them determine the level of student understanding and these can be built into the lesson. They often indicate whether or not students understood the presented material. They are also a way to assess the teaching.

Thumbs Up or Down

At a predetermined point in the lesson or when the teacher is unsure of student understanding, she asks the class to show, by thumbs, if they are with the lesson or not. Thumbs up says, "I've got it." Thumbs down says, "Help please!" and a wiggling thumb means "I'm nearly there but need some extra support." Or thumbs up can mean, "I agree," thumbs down can mean, "I disagree," and a wiggling thumb can mean, "I'm not sure."

Jean teaches in an urban private, special education school. She has six children in her class and has a full time aide. On the Smartboard Jean had projected a series of sentences. Each sentence contained a blank that would be filled in by a vocabulary word. As she displayed each sentence a volunteering student went to the Smartboard and filled in the blank with one of the vocabulary words. She asked the class, by show of thumbs, if the student was correct.

Turn and Talk

This is a strategy that most teachers know, but few use it well. It allows us to pay attention to the student angle in a number of different ways. First, it is an excellent method to use when activating prior knowledge at the beginning of a lesson. Second, it helps to determine if a child's silence is thoughtful or thoughtless. The quiet child who normally does not speak up during whole class discussions is given a chance to speak up and explore with a partner. Additionally, if the teacher raises a question and institutes a turn and talk, students have more time in which to respond. Every teacher has the anxious few whose hands go up immediately after a question is asked. But, using the turn and talk, the teacher allows the quiet student, the ELL, the student who does not process the question immediately, a chance to become engaged and she gets the chance to hear from students who usually do not speak up in whole class conversations. Third, after a strategy lesson is taught, the teacher can initiate a turn and talk to assess the children's understanding as they have a try at the strategy. Finally, before sending the children off to do independent work, the teacher can invite students to teach each other what they are going to do. This allows the teacher to listen for potential problems and for children who might need extra help.

A turn and talk is done with a partner. At a predetermined point in a lesson, the teacher will ask the class to turn to a partner and discuss a specific topic, share an answer to a question, or to make a prediction. As the children are involved in conversation, the teacher moves around and listens in. The teacher will set a time limit for the turn and talk and also have partners share what they talked about.

Betty teaches first grade in a diverse neighborhood. Over 65 percent of the children come from homes where the first language is not English. An ESL instructor pushes in ninety minutes each day. She tries to incorporate opportunity for her students to engage in a lot of talk during the day and uses the turn and talk strategy often.

Betty was reading a story to the children as part of their new unit on birds. She mentioned it was a fiction piece and the children made predictions about the story. Four different types of birds were playing ball and an alligator was carefully watching them. The alligator was getting angrier by the minute. Betty stopped and asked, "Why do you think the alligator is getting angrier by the minute? Turn and talk to your partner about why you think the alligator is getting angrier by the minute." Betty moved around and listened into the conversations. She then said, "We are back in five, four, three, two, one." The class was settled and quiet and ready to share what they had talked about with their partners.

Clearly, Betty's students were aware of the turn and talk method. But it wasn't a routine that Betty used without thought. She wanted to listen to the conversations in order to find out if the children were making logical predictions based on the story they had been listening to and with their knowledge or lack of knowledge about alligators. Based on what she had heard in the turn and talk she could ask the children to share what they had talked about or she could go back into the story, build knowledge about alligators, or proceed with the lesson as planned.

All of these strategies allow the children to be engaged in the meaning-making process. More importantly, they allow the teacher to feel the pulse rate of the children and make modifications in her teaching.

Quick Write

In a quick write, students focus on getting their ideas down quickly without worrying about mechanics. Choose a prompt that sets the focus for the students. It should be one that will allow you to determine students' level of lesson understanding.

For example, Jerry, a fifth grade teacher, gathered his students to the rug to discuss the narrative essay they would be writing. He had prepared a mini lesson in which he modeled the beginning of the literary essay based on some big ideas generated by the class during a previous discussion. Yet, when he looked out the class seemed a bit "dreamy." He began to worry that they would not be able to complete the assignment but wondered where to begin to scaffold the lesson. He flipped the chart paper to the page of notes from the day before where students talked about characters. He gave each student a piece of paper and asked them, "What are the big ideas about character we discussed that you could use in your literary essay?" He allowed them five minutes to write and then asked them to share what they had written. From their responses Jerry knew how to proceed.

Again, this allows the children to be more thoughtful, as writing takes a bit more time than talking. And, it allows the teacher access to children's thinking and understanding, which can inform instruction.

Although quick writes are usually student produced, teachers may use them when recording observation notes on children. A teacher creates a template out of address labels where each label has a child's name or initials in the corner. As the teacher walks around the room observing children's work, she records observational notes on each label. This way she keeps track of who she has observed and a quick look through the notes can yield a pattern. Similar to a focus of a quick write for a student, the teacher also sets a focus for her observation and quick write summary.

Exit Slips

The purpose of exit slips is to engage the students in summarizing their learning. It can be a one sentence summary, an answer to a review question, a question the student has related to the topic of the lesson, a short list of things learned, or the student can set a learning goal for the next class. Many teachers use prompts or stems in order to access information about student understanding. Some examples are:

1. I learned from_____.
2. One question I thought about today's (writing lesson, reading passage, etc.) was_____.
3. One thing I found in today's (writing lesson, reading passage, etc.) was _____.
4. Something I never did before was _____.
5. I'm still wondering about _____.

The possibilities are endless and can be constructed to meet the individual purpose of the lesson. The teacher can assess student learning via the exit slips and plan the next lesson in response to the information on the exit slips.

Gerry teaches third grade in an all-boys parochial school. The parents of the children expect a very traditional program and are very vocal about the academic program in the school. He hesitates to deviate from the prescribed curriculum for his grade. In fact, little time is left for class discussion. Recently he was heard saying, "Hold on! Hold on! I haven't asked a question yet" in response to raised hands. So when the children were leaving class and struggling over homework assignments as an extension of the lesson, Gerry didn't know what to do.

He had a professor who used exit slips in an undergraduate course he had taken at the State University. He thought about trying them with his class. The next day after delivering a lesson on the parts and purpose of a persuasive essay, he asked the class to write a summary of what they had learned. That night while preparing for the next day's language arts class, Gerry read through the notes. He saw a pattern in their summaries and knew the next day he would have to review before asking the class to start their writing. He was glad he didn't just plow on through the original plans frustrating the students, their parents, and him.

In these ways the teacher can be involved in assessment during the lesson rather than waiting until its conclusion, with the exception of exit slips, which still give the teacher important information before moving on with to the next portion of the lesson. The idea is for the students to give feedback, to build it into the structure of the lesson as a necessary and planned part of instruction. Observing, dialoguing, asking for a demonstration or a written response are methods teachers can use to examine students' understanding or to locate the gaps between their knowing and the lesson goals. However, the teacher should always be prepared to alter any lesson based on the information gleaned from an assessment.

The Tool for Practice *Building in Assessment During Instruction* will help you to recognize where in the lesson you could add the suggestions above to help you assess your students' learning during instruction.

Small Group Instruction

Small group lessons, too, provide another opportunity for teachers to learn from students. During small group lessons, teachers can also note students' prior experiences and seeming voids in the knowledge they bring to lessons. This can also be a time to look for patterns of strategies students in the class are employing and strategies students seem to need. Checklists can be developed and employed to help teachers keep track of the skills the students own and which they seem not to own. Let natural curiosity about the children in the classroom guide and direct explicit teaching and areas in which children need practice. The patterns found can be kept in mind when planning whole class instruction. Children benefit from a teacher who listens to them, observes their behaviors and uses the information to inform instruction.

A Tool for Practice *Finding Patterns in Your Notes: On Student Works or From Conferencing* helps you to uncover the patterns in student work that you could use for whole class instruction.

Cheryl works in a public pre-kindergarten class in an urban district. Most of the children in her classroom come from monolingual, Spanish speaking homes. She stated,

> Adjusting to the demands of Reading First Curriculum and a principal that pushes three-year-olds to become conventional readers and writers has been a trying experience for me as a teacher. I have struggled with balancing my beliefs as an educator and the rigorous curriculum that is being forced in my classroom...As I become more comfortable in asserting my own beliefs I have tried to structure my classroom in ideals that value community, authentic learning and joy.

Looking at the Bigger Picture

The ultimate goal is for instruction to lead students to master the state, district, school-wide or personal academic goals that have been set for them. These guide the teachers' efforts in planning, implementing, and assessing our teaching and student learning. How can students' learning be ensured? How can it be determined that students are progressing adequately? How do teachers use their knowledge about student performance to help plan effectively? A Tool for Practice *Thinking Back on Student Learning* (table 7.1) is designed for you to begin reflecting on your lessons and to begin thinking about how student learning became visible to you.

In order to learn from our students we need to discover where the students are in relationship to the curriculum. It means finding their interests and their gaps in understanding and locating the patterns across our class to inform planning and teaching. It means making decisions about the students' experience and the skills they need. It means having purposeful and planned formative assessment throughout our lessons. It means modeling, demonstrating, integrating, and adopting new instruction techniques rather than falling back on old, comfortable ways. Teachers need to expand their ideas of literacy instruction, to broaden the tools and strategies they currently use in order to implement lessons that are responsive to their students' needs. They need opportunities to expand their instructional strategies and practice. And in order to do these, we need to form study teams to provide the support, insight, and ideas we need to be able to teach effectively.

At the Phi Delta Kappa Summit in November of 2008, Thomas Guskey, the keynote speaker, said:

> Maybe best practices depend on where you are, the kind of students you're teaching, the kinds of communities in which they live, the cultural background they bring to school. Those things really need to be built in, because if what's effective depends on the kinds of students who are in front of you, then we have to prepare our teacher candidates to really be familiar with those kids, what they're facing, and how they can be effective with them. (Young, 2009, p. 439)

We have provided tools at the end of the chapter for the teacher to use in an effort to plan for the students in her classroom in a responsive manner. It is our hope new habits will grow from these tools to increase knowledge about student learning in order to adjust teaching.

Tool for Practice: *Building in Assessment during Instruction*

There are many opportunities to assess student learning during a lesson. Some of the suggested methods were Thumbs Up, Thumbs Down, Turn and Talk, Quick Writes, Graphic Organizers, and Exit Slips. There are others you might be comfortable with or might occasionally use with your students. The idea is to plan formative assessment while you are planning your lesson. Use the sample lesson plan to add some of the suggested methods for assessing student.

Then, carefully go over your lesson procedures. Identify parts of the lesson students need to grasp before being asked to accomplish a task, work independently, or move on to more difficult concepts. By building in an assessment at those points you can monitor student learning. You will discover immediately if you need to stay with that portion of the lesson to support student learning or if you can move on with the lesson as planned.

Here is a common lesson plan. Fill in the boxes along the side with suggestions from the chapter for assessing students during instruction.

Subject: Language arts

Topic: Identification of nouns

Objective: At the end of the lesson students will be able to:

1. Identify nouns in a sentence.
2. Define what a noun is.
3. Complete a word sort by categorizing words as nouns or not nouns.

Procedures:

1. Invite the children to the rug.
2. Show them the chart paper on which is ⟵
   written:

   "Writers start identifying the parts of
   a complete sentence by picking out the
   nouns in a sentence."
3. Introduce lesson with "hook." Explain
   to the children how much I like cars,
   how I enjoy studying them, and so on.
   In order for a car to work it's made of
   many different parts on the inside. Can
   you name some of these parts?
4. Show pictures of car parts. Ask ⟵
   students to identify them.
5. I get really excited talking about cars
   and I'm excited to talk about nouns.
   In the same way that cars are made up
   of different parts that work together,
   sentences are made up of parts that all
   work together.
6. Explain to students that nouns are very
   important parts to a sentence.
7. Explain that a noun is a person, place or ⟵
   thing. Write definition on chart paper.
   Have student choral read.
8. Using index cards and a think aloud ⟵
   strategy, model a word sort using two
   categories: Noun and Not a Noun.
9. Have a few students volunteer to suggest
   a category for a word.
10. Show paragraph previously written ⟵
    on chart paper. Ask for volunteering
    students to circle nouns they find in
    paragraph.
11. Send children back to seats.
12. Distribute worksheet for word sort
    identical as one modeled.
13. Instruct students to sort words into ⟵
    appropriate categories.
14. When finished, ask them to write a par-
    agraph about their favorite thing to do
    on weekends. Then circle all the nouns
    the used.
15. Have a few students read their ⟵
    paragraphs and share how they identified
    which words were nouns.

Table 7.1   Thinking back on student learning

Lesson I taught	Students' reactions	Reflection
Describe the lesson List the various component to the lesson in this column	For each component of the lesson, record the students' reactions. You could list the questions they asked, their responses to assessments you gave them, notes you collected	What do the reactions mean about students' learning? What seemed to work? What caused confusion or problems? How can you adjust the lesson for future learning?

Tool for Practice: *Thinking Back on Student Learning*

By breaking down your lesson into parts, you have the opportunity to see what went well and what did not go as well as you had predicted. In this way you have the opportunity to adjust your future planning and teaching.

Suggested Resources

Cunningham, P.M., & Allington, R.L. (2007). *Classrooms that work: They can all read and write* (4th ed.). Boston: Pearson.

Harvey, S., & Goudivs, A. (2000). *Strategies that work: Teaching comprehension to enhance understanding.* Portland, ME: Stenhouse Publishers.

More Teaching Ideas is a website where you can get and post ideas www.teachingideas. co.uk.

References

Allington, R.L., & Cunningham, P.M. (2007). *Schools that work: Where all children read and write.* Boston: Allyn and Bacon.

Altwerger, B., Arya, P., Jin, L., Jordan, N.L., Laster, B., Martens, P., et al. (2004). When research and mandates collide: The challenges and dilemmas of teacher education in the era of NCLB. *English Education, 36*(2), 119–133.

Anderson, L.W. (2009). Upper elementary grades bear the brunt of accountability. *Phi Delta Kappan, 90*(6), 413–418.

Bean, R.M., Cassidy, J., Grumet, J.E., Sheldon, D.S., & Wallis, S.R. (2002). What do reading specialists do? Results from a national survey. *The Reading Teacher, 55*(8), 736–744.

Cunningham, P.M., & Allington, R.L. (2007). *Classrooms that work: They can all read and write* (4th ed.). Boston: Pearson.

Darling-Hammond, L. (1997). *The right to learn: A blueprint for creating schools that work.* San Francisco: Jossey-Bass.

———. (2007). Teacher learning that supports student learning. In B.Z. Presseisen (ed.), *Teaching for Intelligence* (2nd ed.; pp. 91–100). Thousand Oaks, CA: Corwin Press.

Davis, M.M., Konopak, B.C., & Readance, J.E. (1993). An investigation of two chapter 1 teachers' beliefs about reading and instructional practices. *Reading Research and Instruction, 33*(2), 105–118.

Dearman, C.C., & Alber, S.R. (2005). The changing face of education: Teachers cope with challenges through collaboration and reflective study. *The Reading Teacher, 58*(4), 634–640.

Dole, J.A. (2004). The changing role of the reading specialist in school reform. *The Reading Teacher, 57*(5), 462–471.

Fullan, M., & Hargreaves, A. (1996). *What's worth fighting for in your school.* New York: Teachers College Press.

Fullan, M.G. (2003). *Changing forces with a vengeance.* New York: RoutledgeFalmer.

Greenwood, C.R., & Maheady, L. (2001). Are future teachers aware of the gap between research and practice and what should they know? *Teacher Education and Special Education, 24*(4), 333–347.

Heritage, M. (2007). Formative assessment: What do teachers need to know and do? *Phi Delta Kappan, 89*(2), 140–145.

Hollingsworth, S. (1989). Prior beliefs and cognition change in learning to teach. *American Educational Research Journal, 26*(2), 160–189.

Ivey, G. (1999). Reflections on struggling middle school readers. *Journal of Adolescent and Adult Literacy, 42*(5), 372–381.

Lipson, M., & Wixson, K. (1997). *Assessment and instruction of reading and writing disability: An interactive approach* (2nd ed.). New York: Longman.

Lortie, D.C. (1975). *School teacher: A sociological study.* Chicago: University of Chicago Press.

Luke, A. (1994). *The social construction of literacy in the primary school.* South Melbourne, Australia: MacMillan Education.

McGill-Franzen, A., & Allington, R.L. (2005). The gridlock of low reading achievement: Perspectives on practice and policy. In Zhihui Fan (ed.), *Literacy teaching and learning: Current issues and trends* (pp. 173–183). Upper Saddle River, NJ: Pearson Merrill Prentice Hall.

Roller, C.M. (1996). *Variability not disability: Struggling readers in a workshop classroom.* Newark, DE: International Reading Association.

Routman, R. (2000). *Conversations.* Portsmouth, NH: Heinemann.

Theriot, S., & Tice, T.C. (2009). Teachers' knowledge development and change: Untangling beliefs and practices. *Literacy Research and Instruction, 48*(1), 65–75.

Toll, C.A., Nierstheimer, S.L., Lenski, S.D., & Kolloff, P.B. (2004). Washing our students clean: Internal conflicts in response to preservice teachers' beliefs and practices. *Journal of Teaching Education, 55*(2), 164–176.

Young, E. (2009). What makes a great teacher? *Phi Delta Kappan, 90*(6), 438–439.

8

Learning from Parents and Families

Parents are not involved [in my classroom]. I don't really know how to involve them. I feel intimidated by them.

—(Kindergarten teacher at a private school in Manhattan)

They could come in during Open House, which happens a few times a year.

—(Second grade teacher at a public school in Queens)

They can contact me whenever I have a prep, by making an appointment first.

—(Fourth grade teacher at a public school in Manhattan)

Challenge and Opportunity

These comments were made by teachers when we asked them to describe ways in which they involved parents. This perception of parental involvement shared by the teachers mentioned in the epigraph is all too common. For a variety of reasons they are more comfortable keeping parents out of their classrooms; some worrying that they might lose control, fearing that parents might try to micromanage the work they do. Some teachers do not trust parental judgment and feel parents do not have anything worthwhile to add to the daily operations of academic life. And, some teachers do not know the benefits of parental involvement; they are just mirroring the culture of the school toward parental and family involvement. With little formal education on working with parents, few professional development days devoted to home-school partnerships, and school cultures void of well-planned family-school partnership events or family initiatives, teachers are left with little support and few resources on which to fall back when trying to involve parents in their own classrooms.

The teachers mentioned earlier, like many teachers, expect parents to come to school for open school nights, parent-teacher conferences, and other school sponsored events and often interpret low attendance at these functions as lack of interest on the part of the parents. Teachers want parents to keep them abreast of problems, yet frequently make it difficult for parents to contact them. Parents may not feel invited or welcomed by the teacher. The school may say that parents are important but the school culture may actually discourage them from attending events. Schools failing to recognize the changing demands on families in dual career homes and parents' failure to recognize the demands on teachers interfere with the ability to establish home-school relationships.

There is growing evidence that family involvement helps to promote student achievement and school quality. Yet family involvement will only succeed if teachers are adequately prepared for home-school partnerships. For many teachers it means learning how to work with parents in new and often creative ways. For others, it means learning how to begin the work with families of the children in their classrooms. The growing diversity and changing demographics of the present day classroom can further complicate the development of partnerships. Efforts begin when teachers recognize that parents want to do the best for their children and can make significant contributions to their children's education (Shartrand et al., 1997).

When parents and teachers work together, a single important message is sent to children that says, "We are working together to ensure your academic success." It is important for teachers to push the limits of their own level of comfort in working with parents and to identify ways in which parents feel welcome and a part of the classroom community because the benefits are great! Many of our beginning and experienced teachers have had to work through their own discomfort when working with parents. As they have had opportunities to become literacy coaches, they have had new challenges to overcome. We can learn a great deal from the stories they have shared with us and from the efforts they have made in fostering home-school partnerships.

Story from the Field

Mrs. Holliday, an experienced teacher, was welcoming her class back from spring vacation. The children were chatting among themselves, unpacking their backpacks, registering their lunch choices, or putting their lunch boxes into the lunch bin. The room had a healthy, exciting buzz to it and Mrs. Holliday was looking forward to a fresh start after a week's vacation.

Children were bringing her notes from home: who needed to leave early for the dentist or was going to go home for lunch. She quickly opened the notes to plan for the accommodations she would need to make. Then she opened the note from Ian's mom.

Dear Mrs. Holliday,

I have noticed that Ian's work has become sloppy over the last few weeks. The worksheets he is bringing home has shown some careless mistakes. When I asked him about it, he told me he was getting bored.

At our parent-teacher conference in November, you asked if you could work with Ian and a few of the other children on special assignments and to give them more challenging opportunities. Would you please update me on how that is going and what they are doing.

Also, specifically, how are you going to keep Ian engaged until the end of June?

Thanks so much.

Darlene

Mrs. Holliday looked up and tried to organize the children for the morning routines and then to the math lesson she had prepared. She felt dizzy, angry, and absolutely overwhelmed. When, in an already busy day, would she find the time to answer Ian's mom? Should she call her? No, that may just stir up even more distress. Should she go to her principal? No, that may send the message that she was not competent enough to deal with a parent. Write Ian's mom a note? She didn't have time to think about it now.

Research and Reflection

The importance of creating a connection between the home and school cannot be underestimated. Parental involvement and the partnership between families and school are seen as keys to student achievement (e.g., Briggs et al., 1997; Lazar et al., 1999; Martin & Hagan-Burke, 2002; Shockley et al., 1995). Urban teachers, faced with increasingly diverse populations, have many more challenges when trying to form partnerships with parents. Teachers now face an increasing range of student needs, more culturally diverse populations, and host many students who speak English as a second language. According to Villegas and Lucas (2002) to teach successfully in a culturally diverse setting, teachers must see all students, including children who are poor, marginalized, and speakers of languages other than English, as learners who already know a great deal and who have experiences, concepts, and languages that can be enhanced and expanded to help them become more proficient learners. They continue, "Teachers' attitudes toward students significantly shape the expectations they hold for student learning, their treatment of students, and what students ultimately learn" (p. 23).

Lareau (1989) found that parents in upper and middle class homes actively communicated with their children's teachers, saw education as a shared responsibility, and sought strong involvement. Parents from low-income, working class homes tended to keep their distance from the schools and teachers interpreted their low attendance at school functions as lack of interest in their children's learning (Paratore et al., 1995). Teachers are often uncomfortable acknowledging these class issues so it is especially important to look beyond initial reactions and interpretations of parental behaviors to engage parents. In order to build mutual trust and respect, teachers in urban schools need to take a more proactive role in ways to effectively communicate with families.

Knowing that parental involvement is important and putting that knowledge into practice is often difficult for teachers. A number of studies (e.g., see Isenberg & Jalongo, 1997; Shartrand et al., 1997) suggest that teachers are not prepared to work with families nor are they prepared to design and implement effective methods to communicate with families of their students. Many teachers would like to create partnerships with parents but have not received adequate preparation to do so. They have not been taught strategies to engage families in their children's school life socially or academically. They often do not know how to keep in touch with families and the importance of communicating regularly, to reach out to show they value all children in their classroom. Displaying student work outside the classroom, planning programs and activities for families, special workshops, developing learning kits to send home, inviting parents to help plant a garden, organize a sports equipment swap, sew costumes, read to the class, help cut out templates for class projects are just a few of the ways parents can be invited to participate and most teachers are not prepared to create partnerships is this way. Hoover-Dempsey and Sandler (1997) found that parents are more motivated to support their children's learning when they receive clear invitations and support from teachers. They suggest teachers give parents a job description of parental involvement.

Moll and his colleagues (1992) believe that children have a background of learning experiences and a network of people who support various kinds of learning; that children have "funds of knowledge" that teachers rarely draw upon for use inside the classroom. "In developing our partnership with families, we are not trying to impose our vision of literacy but to develop relationships with families where we could learn about what already existed in the families and connect with that literacy classroom community" (p. 94). To do this, teachers need to see themselves as learners and be willing to put themselves in a collaborative position with parents. This can be initially uncomfortable for many teachers but necessary to establish real partnerships.

The difference between the literacy activities in the home and the school has been well documented. When we can widen our definition of literacy to include out of school literacy activities, we can deeply engage children and their families. Although homes contain rich literacy practices, they are often incongruent with the practices of the school. A person considered literate in one culture may be considered illiterate when judged by the standards of another culture because adults have always schooled children in some manner to assist them in becoming literate in the ways and traditions of their own cultures" (DeBruin-Parecki & Krol-Sinclair, 2003, p. 1). How parents share literacy events with their children needs to inform classroom literacy instruction (Auerbach, 1995; Baker et al., 1996; Morrow et al., 1993; Paratore, 2002). It provides an entry point for many teachers when working with children of cultures other than their own.

The literacy events of the home provide a strong foundation for later learning yet many parents do not realize the simple, everyday activities they do at home are important. As teachers we need to tell parents that creating shopping lists, cooking together following recipes, drawing and talking about their pictures, noticing print, going through the mail, and other parts of their daily routines as well as reading books and visiting the library are important to their children's literacy development. In fact in similar studies conducted by Purcell-Gates (1996) and Taylor and Dorsey-Gaines (1988) it was found that children who were exposed to a great variety of type and frequency of literacy tasks in their homes scored high on print tasks in school settings and were generally successful in school. Time spent with computers, television, reading baseball cards, keeping track of scores of favorite teams, and conversations around daily events are important connections to home literacy. Teachers need to cultivate and acknowledge the literacy strengths of each family.

Revisiting Mrs. Holliday

Mrs. Holliday dismissed the class to lunch. She thought about how she should respond to the letter she received from Ian's mom. She sat at her desk, sandwich in hand, and decided to write a note in return.

Hi Darlene,

Thank you for your note. I've also noticed some careless work from Ian and spoke to him briefly. 4th grade is a hard year socially, especially towards the end of the year. He has taken up a friendship with Kevin. Many times both are distracted because of talking and being silly (normal boy behavior). I know that was the case in grammar...Ian is always enthusiastic during lessons and volunteers constantly. He is also very friendly and well liked. At this age and time of year I see the social side of life become more important than grades with 4th graders. This phase usually passes.

Aside from grammar skills, (which lessons are usually dry) I will briefly describe what work Ian has worked on in each subject and will work on till June.

Mrs. Holliday continued the note for a total of six pages outlining every subject area, what the class had been learning, and the activities she had planned for the remainder of the school year. Had she built a relationship with the parents of children in her classroom, especially Ian's mom, she would not have had to spend so much time responding to Darlene's concerns. Mrs. Holliday, like many teachers, would hope to implement strategies that promote home-school connections. The following suggestions are based on work we have done with teachers, like Mrs. Holliday, to develop their abilities to foster parental involvement. They are designed to help you find an entry point, a beginning in your work with parents to create a home-school connection. Some of the suggestions here are designed as a generic way to move you out of your comfort zone and to begin to build partnerships with parents. However, our ultimate goal is for you to enhance literacy learning by extending school curricula to family routines, elicit information from parents about students' home literacy practices, and to become collaborators and mutual supporters for their children's academic success.

Strategies and Tools

Nearly all families have some forms of literacy and language events in their daily routines. They can be different from those teachers recognize or ones that are found in the routines of school. Surveys asking parents about language events practiced at home is a good entry point. By seeking to learn home practices, the teacher can learn about and integrate parents' existing knowledge and literacy and language practices into the classroom. This acknowledgment of home literacy events puts value on them and initially invites parents to form a partnership.

Also surveys asking how families would like to be involved and what they need could be a part of that initial survey. It is a more generic way to involve parents until you are comfortable moving the relationship forward. There is no "one way" in which parents can be involved just as there is no one way literacy is practiced in each home. Our vision must be broad and we must acknowledge that parents want to be involved and they want their children to do well. Engaging families in their children's school life in a variety of ways so all families feel included and respected may be uncomfortable at first. To help teachers through this difficulty, we have organized a step-by-step guide for the teacher in planning for parental and family involvement. This would allow for teachers to reach their own comfort zone.

We have included two sample surveys we developed along with our students. One titled "What Are Your Family Routines?" asks for common literacy practices found in the home and the other is a more generic Sample Parent Survey to start the home-school connection. Use them as templates, add or delete items that best reflect ways you want to involve the families of your students. Remember to be as inclusive as possible, thinking about the schedules and talents the parents of the children in your classroom possess. Allow parents the opportunity to suggest ways for developing partnerships as well.

Classroom Activities to Promote
Home-School Connections

It is important for teachers to draw upon family literacy practices and view them as important resources. Knowing and understanding the cultural backgrounds of students can be used in discussions and book readings of specific holidays. Family memories and stories can be brought in and shared (Winston, 1997). One way is to collect family stories and create class big books. For example, families can send in favorite recipes or traditional recipes from their backgrounds and take photographs of how they prepare and serve the food in their homes. Children and parents can write captions for the photographs together. These books can be shared in class but can also be borrowed each night by individual students to share with their families. Another way is to have family members come in and share their favorite or culturally traditional stories with the class on regularly scheduled days. What a rich way to enhance the literacy practices in your classroom while recognizing those in the home!

Regularly planned school events can be a good way to begin efforts of making families feel comfortable and connected to the school. At the end of the chapter you will find samples that may help you plan some at your school. Included is the invitation you might send to the families of children in your class or grade to come to a Family Timeline Night. In addition, we have included how to organize for the Family Timeline Night. They can be used as templates for any themed night you plan for families.

During Back to School Night teachers can make it clear what parents can do at home and at school to promote learning. If a teacher expects parents to check homework every night, parents need to know what to look for, how to do it, and how much help to give the children. Should the parents correct the mistakes? Or would the teacher rather see where the child still needs help? How long should homework take each night? Parents should leave Back to School Night with a clear understanding of what the children will be learning and doing in class, the Standards and how the class work promotes those Standards, and what parents can do at home to support the classroom work. Parents need to know what good work at this grade looks like and how it is evaluated. Show parents how the room is set up, explain learning centers, word wall, the author's chair, classroom library, and computer center. Discuss how the room set up helps to meet reading and writing learning goals.

Bulletin board exhibits of student work with explanations of the purpose of the work and the Literacy Standards met by the work can be very informative and keep parents abreast of learning in the classroom. Newsletters can include some of the work samples with similar or more in-depth explanations.

Special workshops to show how families can help their children at home and in response to what families feel they need to know can be scheduled throughout the school year. Topics such as how to read to your children, select books, and how the report card is connected to the school, district or State Standards can help parents stay informed (Tuten, 2007). Remember that most parents are curious to see what goes on during literacy instruction. Consider showing short video clips or photographs of you and your students at work. Workshops can also be videotaped for parents who may not be able to attend. Parents can then borrow these tapes to watch at home.

It is common to invite parents to share their experience and knowledge within the classroom, so that parents are not kept at the margins of their children's learning but feel involved. Organizing career days or career nights, depending upon the work schedules of parents, can tie into units on community workers. When parents cannot attend, invite them to take photographs of their work environment and to write captions for the photographs to be shared by their children during career days. The home-school-community connection can be established and the children have the opportunity to learn more about their parents' jobs. When possible, connect books used during read-alouds to the various jobs parents have. Many parents do not have the time to spend in classroom so inviting them to send in traditional stories, family scrap books, and photographs are a way to keep them included and reaffirm their value.

Helping Parents Help at Home

One of the teacher's most important tasks is to help parents understand how children become readers and writers (Enz, 1995). As discussed earlier, acknowledging the importance of home literacy events and helping parents identify the literacy practices engaged in at home as a factor in school success is one avenue to parental understanding. When we asked our teachers what they believed the parental role should be in helping children gain academic success, our teachers made very general comments such as, "every way possible" or "parents have a huge role." Yet, many parents do not know how to help their children or the level of help the teacher expects. Nor can they come into regularly scheduled Back-to-School Nights to ask questions and understand the teachers' expectations. Parents need clearly defined roles the teacher expects them to play. By knowing what their children are studying in school, parents are in a better position to talk about the books they are reading, themes they are studying, and the progress of their writing. For example, knowing that the class will be reading a book from the Junie B. Jones series by Barbara Park and having specific strategies for helping their child, parents have the opportunity to become more involved with their children's learning.

Parents need to know how to share books with children. This is a perfect topic for a Family Night or Day Time Workshop with parents of young children. Books along with suggestions for conversations around them can be put into a book baggy when sent home. Audiotapes and tape players for students to use at home may help those whose first language is not English or whose parents are not literate. A journal can accompany the book. Family reactions to the books can be written by the student or other family member and brought into class. Paul, a first grade teacher in an elementary school located in a densely populated, predominantly immigrant neighborhood, taught his students how to read aloud with their parents. He gave them the following instructions:

- Find a book you would like to read. It could be one at home or one you have brought home from school in your backpack.
- Sit next to your mom or dad or grandma or grandpa or auntie or whoever is home and put the book between you.
- Take your parents' arm (or the arm of whomever is reading with you) and put it around you so that it rests on your shoulder. They are your reading partner.
- Open the book and have your reading partner read it to you. Or, you read a page and have your reading partner read a page. Or, if your reading partner can't read, you read it to your reading partner as best you can.

Many parents do not know that sharing books with their children is important. Their culture may be one with a strong oral tradition. There may not be books in the home that are appropriate for children. Paul recognizes this and gives instructions to the children so they can conduct their own read-alouds.

Parents must have access to the literacy practices and literacy related activities and materials to enhance their children's progress (McCarthey, 2000). When teachers give parents directions and specific suggestions for supporting their children's literacy development at home, parents are more likely to be involved (Epstein, 2001).

Monthly newsletters help teachers proactively communicate with parents about the literacy events in school. A newsletter can describe literature circles, author studies, classroom practices regarding independent reading, word walls, or spelling. Specific books, home activities, websites, television programs, museums, or other neighborhood paces of interest that coordinate with topics children are learning in the classroom help to create a home-school connection (Jensen, 2006). There is a template at the end of this chapter that you can use for designing your own newsletter.

When a new child enters the classroom, teachers often assign a classmate to welcome the child and to explain classroom routines and to help the child feel comfortable. Parents, new to the school, can also feel uncomfortable and at a loss. The teacher can ask a class parent to reach out to the family of the new child in class to help them feel more comfortable with the new school.

One of our teachers invites parents to help by sending construction paper and templates of shapes she needs cut for lessons. The parents cut the shapes and send them back into school via their child's backpack. Sometimes, teachers need to be creative in ways to have parents connected to the classroom.

Parents are motivated to be involved when they feel welcome and respected. Teachers need to reach out to parents, to invite them to be involved. Table 8.1 offers suggestions for reaching out to parents. To discover your comfort level in working with parents and for suggestions of how to take the next step for creating a stronger home-school partnership, take the Teacher Survey at the end of this chapter. It is the teachers' responsibility to make the first step in creating a home-school connection by telling parents how they can stay in touch. As one of our teachers stated, "I see now that I have to become more dependent on taking the initiative of getting the parents involved and become the leader, than just assuming that they are going to do it on their own." Another of our teachers said, "As a

Table 8.1   Strategies to foster parental involvement

What to do	Ideas for implementation	Benefits
Learn who the parents are and their expectations	• Survey parents asking about their literacy practices at home • Survey parents about how they would like to be involved • Ask how parents would like to be involved at Back to School Nights and Parent Teacher Conferences	• Acknowledges parents role as important • Helps parents feel respected
Draw on family practices and cultural backgrounds	• Ask parents to share traditional stories • Have students share family scrap books and photographs • Collect family recipes and photographs of food preparation	• Family traditions and cultures are valued • Children learn about their classmates • Information can inform future instruction
Introduce parents to classroom and classroom practices	• Back-to-School Night should include a tour of the classroom explaining centers, word wall, etc. • Discuss with parents how the room set up helps to meet instructional goals • Take photographs or videotape clips of classroom and practices • Bulletin Board exhibits of student work	• Enables parents to "see" what goes on during their child's day and breaks down the separation between home and school • Keeps parents abreast of daily routines • Parents understand instructional goals
Help parents develop ways to support their children's learning at home	• Monthly newsletters • Workshops • Family reading nights • Book baggies with journals • Parent-to-parent connections	• Parents feel involved in child's learning • Richer conversations about school • More directed and engaged homework help

teacher I must realize that parents do not deliberately choose to take a passive role in their child's education. Sometimes they do not know where to begin. The home-school connection requires persistence and a bit of 'stepping out of the box.' It is not easy."

What are Your Family Routines?

Every family has different routines and family rituals. They are the things that make each family special and unite a family together. I am interested in finding out about the special ways you share reading, writing, listening, and speaking in your home. The check list here (table 8.2) lists some of the ways many families share literacy events either daily or at special times. Please use this check list to help me learn about your family routines. I have left room at the bottom of the chart to include other ways your family is special.

Table 8.2   Discovering family literacy routines

Event	Every day	Occasionally
Talk about our events in our day		
Make grocery lists		
Write notes to each other		
Send cards for special events		
Talk about school		
Write letters		
Send and read email		
Text message each other or friends		
Read a newspaper		
Read books		
Tell stories		
Sing songs		
Pray		
Google or use other search engines		
Talk about TV shows		
Chat on the phone		
Use recipes		
Read food labels		
Read store signs		
Read menus		
Pay bills		

Sample Parent Survey

I would like to find out ways you can share and contribute to the life of our classroom. Your involvement is important to me and to the children. Please fill out the survey and return it to me as soon as possible. Thank you. I look forward to working with you this year.

Name_____ Child's Name

Phone Numbers(s):

        Home_____

        Work_____

        Cell_____

Email Address:_____

What is the best time for you to come to school?

_____Morning _____Afternoon _____Evening _____Weekends

Is there anything that prevents you from becoming involved that I would be able to help you with?

_____Transportation _____Child care

_____Other:_____

I am willing to be involved in the following way(s):

_____Make phone calls

_____Assist with class projects

_____Attend field trips

_____Help arrange classroom events

_____Read to the class

_____Speak on a special topic

_____Go to the library to borrow books for topics we are studying

_____Cut our templates at home for class projects in school

Many parents have special talents. What are your special talents that you would be willing to share with us?

_____

_____

_____

I would like to host family nights and offer parent workshops this year. Are there any topics you would like to see addressed? (i.e., State Standards,

Grade Curriculum, Tips for Homework Help, Choosing Books for Your Child)

Sample Family Workshop Materials

Date

Dear Family,

You are invited to:

              4th Grade Family Time Line Night

Who: 4th graders and their families

When: Date

Time: Time of the event

Where: Location of the event

Why: The children have been studying New Jersey history this year. As a final project they will be creating a time line of all the historical events they have studied about New Jersey this year and add to it the events they learned last year about the history of our town.

Before we create our New Jersey Time Line, we want to make our own family time lines and share and learn of the many families in our class.

Bring in some photographs or drawings of your family's special events to put on your time line.

Snacks and drinks with be provided.

Return this form before (date).

_____Yes! My family and I will attend Family Time Line night on (date)

Student's Name_____

Number of people attending_____

_____Sorry, we can't make it.

Organizing Family Time Line Night

Before the night:

- Make your own family time line to use as a model during the evening.
- Help organize snacks and drinks.

- Arrange tables and chairs into small work groups.
- Provide long sheets of paper, rulers, markers, glue sticks, crayons, and other materials families may need to construct their time lines.

At the event:

- Explain the purpose of the event and the standards or curriculum mandates the time line creation may meet.
- Explain the plans for the evening.
- Show your model.
- Help families work on their time lines.
- Invite families to share in small groups if they like.
- Take photographs of the families working together.

Hang their time lines on a bulletin board and post photographs from the event around it.

Provide an evaluation form for the event.

Thank the families for coming. You may want to give them suggested websites or books they can use at home that extends the learning of families.

For families unable to attend, send home a note, tell them you missed them and tell them about the event. If you gave a handout about suggested websites or books, include it with the note. Invite them to create their own time line at home and include it on the bulletin board. Invite them to look at the time lines the class created.

Notes on Newsletters

Title of Your Newsletter
Date of your Newsletter

Dates to Remember

Greeting from the Teacher

This short column should be written in a warm and respectful manner. It sets the tone for the rest of the newsletter. Alert them to upcoming events, thank them for their support and help for certain projects, or tell them what to expect this month.

Dates for school photos, field trips, back to school nights, conferences, school holidays for the month should be listed here.

If something extraordinary is happening the next month you might want to include it in this list.

You can bullet these items or list them one under the other.

This Month's Websites

Share the school website or your own classroom website address here. You might also list the website for the location of a class trip, sites you found that would enhance instruction and that parents and children could navigate together. For example, when doing an author study, list the author's website or if you are studying the planets websites for further investigation could be listed.

> We Need Your Help This Month.

This is your chance to get the parents involved in the life of your classroom. Perhaps you are studying time lines in social studies and math. You can invite families to create their own family time line to be shared in class. Give them the opportunity to include photographs or to bring in objects.

Look What We Did!

Share the work that your students completed during the previous month, especially work that you had mentioned in the newsletter. Parents are more apt to read the newsletter and children are more apt to share the newsletter with their parents if their work is included. Scan it in, type in comments made by the children, include photographs of bulletin boards with their work. In this way the newsletter becomes a vehicle for communication.

Units of Study

Using subheadings, share with the parents the topics and activities the children in your class will be involved in during the month. What books will they be reading? Will they be participating in literature circles? If so, tell the parents about them. What will they be learning in science? Mathematics? Social Studies? Art? What special skills or strategies will they be mastering this month? What are your goals?

Drawing Parents Out

In this section of the newsletter you want to draw out parents with knowledge or experience in the topics or themes being studied. By inviting those parents into the classroom to share their knowledge and experiences, you acknowledge them as valuable contributors to their children's learning.

Common family practices that support the topics and themes can be mentioned here. For example, when studying fairy tales, invite parents to share the ones they tell or read at home with the class.

Invite Feedback

Find a way for feedback. This could be a simple note or family journal entry about a parent-child activity tried out at home, a book they shared, a trip they took, or a response to an invitation made by you to help in the classroom.

Survey for Teachers

The following tool will help teachers determine their level of comfort when working with parents. Suggestions relating to each level of comfort will help teachers make the next step to creating home-school connections.

The following multiple choice questions are designed to help you determine your comfort level of parental involvement. Choose the response that *best* describes your current practice.

1. My school's policy about parental involvement requires that I
   a. communicate monthly with parents of the children in my classroom.
   b. communicate monthly only with parents of children who are struggling in my classroom.
   c. communicate with parents during school sponsored activities such as Parent Teacher Conferences and Back to School Nights and when a child is having a specific problem.
   d. communicate only with parents when a child is having a specific problem.

2. Which best matches my current practice?
   a. When we are involved in an author study or thematic unit, I send home a variety of books, websites, learning activities, and suggestions for families to do at home and discuss how the work in school and at home meets current learning standards.
   b. Part of my monthly newsletter discusses the areas we are studying in literacy and suggests books or websites that may be of further interest.
   c. I alert parents to the areas we are studying in school and ask them to make sure their children complete all assignments.
   d. The parents of the children in my classroom do not have the time or the interest to help and I would rather not have them involved in what I am teaching.

3. Parents understand the purpose of the work their children do in my class by
   a. reading the monthly newsletters I send home explaining the work.
   b. looking at the bulletin board exhibits outside my classroom explaining the purpose of the work.
   c. visiting the classroom on school sponsored open school days.
   d. how well their children do on graded work.

4. When working with parents I
   a. explain ways they can monitor their children's work and build their skills.
   b. tell them to have their children follow my directions on homework and other assignments.
   c. leave it up to the parents to help with homework in the way they know best.
   d. cannot count on parents caring or helping their children with their homework.

5. When I select a book for my class to read in literature circles or book clubs,
   a. the parents are told ahead of time and have the opportunity to join a parent book club and discuss the book before their children read it in class.
   b. the parents are told about the book in the monthly newsletter I send home.
   c. the parents learn about the book from their children's work related to the book.
   d. the parents learn about the book if their children decide to share their experiences and discussions about the book.

6. The activities for parent and family involvement at my school include:
   a. A plan to engage families of the children's in their academic life *and* social activities to help make families feel comfortable and connected to the school.
   b. Social activities to help make families feel comfortable and connected to the school.
   c. Back to School Night, Parent-Teacher Conferences, other whole school sponsored once a year events.
   d. Parent-Teacher Conferences.

7. As a teacher,
   a. I instruct parents in school-based literacy and seek to learn about and integrate parents' existing knowledge and resources into school curricula.
   b. I instruct parents in school-based literacy and offer opportunities for parents to share their knowledge on specific topics I have selected.
   c. I instruct parents in school-based literacy practices.
   d. I instruct parents in homework routines.

8. Parents of children in my classroom learn about classroom activities mostly by
   a. specially deigned workshops I offer parents.
   b. weekly or monthly newsletters I send home.
   c. talking to their children about what occurs in school.
   d. looking at the work their children bring home.

9. My idea of parental involvement is
   a. a partnership where families and the school work closely together to ensure student success.
   b. parents and families being involved in planning and getting people out only for school sponsored activities.
   c. parents volunteering when help is needed and working with their children at home.
   d. parents helping their children at home.

10. At Back to School Night, parents of children in my class
   a. get a clear idea of what their children will be learning and doing in my class by introducing them to their textbooks, the room set up and how it helps to meet the standards for student work, possible field trips, and homework expectations.
   b. are introduced to how the room is set up and how it encourages student learning (word wall, library, etc.).
   c. are told what we will be studying, what my expectations for homework are, and possible field trips we will be taking.
   d. are told what we will be studying and are invited to look at their children's text books.

11. What is the level of parental involvement at your school?
   a. Parents play an active role in forming policy and curriculum.
   b. Parents are active volunteers, helping the school raise money.
   c. Parent groups help implement an agenda set by the principal.
   d. Parents help for specific projects when asked by the principal.

12. In my classroom,
   a. I work with parents by sending home learning packets, educational games, and videos linked to what the children are studying in school.
   b. I make it clear what parents can do at home to promote learning by making suggestions linked to what the children are studying in school.
   c. I expect parents to work with their children on topics we are studying in school.
   d. I do not expect parents to be familiar with the topics we are studying in school.

13. Which of the following best describes your attitude about parental involvement in your classroom?
    a. I believe that every family has something to contribute so I send home a survey at the beginning of the year to find ways we can work together.
    b. I believe parents should help in my classroom so I often send home letters requesting help for certain projects.
    c. I ask parents to help with class parties and to chaperone on class trips.
    d. I do not want the parents in my classroom. I find them a disruption.

14. The parents of children in my classroom
    a. know the value of reading aloud to their children and know how to interact with their children around books from workshops I have conducted.
    b. know the value of reading aloud to their children and, I hope, follow suggestions I made at Back to School Night.
    c. read aloud to their children each night from a selection of books I send home.
    d. should read to their children every night.

15. Home literacy events
    a. should be incorporated into the life of my classroom.
    b. are important to know about so I understand the students' home lives.
    c. are those reading and writing activities children do at home.
    d. are not present in all my students' homes.

16. Which statement below best matches your thinking? As a teacher,
    a. it is my responsibility to help parents understand that their home literacy events play an important role in helping their children become successful readers and writers and must build parents' knowledge of how to support their children's literacy development in school and in the home.
    b. it is my responsibility to help parents realize how their simple, everyday interactions with their children establish a foundation for literacy.
    c. it is my responsibility to instruct parents on how to implement my literacy related activities at home.
    d. it is not my responsibility to instruct parents on classroom routines and expectations. I do not want to interfere with their lives.

For each of the following questions, circle all those responses you feel best matches your current practice.

17. In order to help me understand my students,
    a. students bring in a portfolio of samples of literacy events that are part of their family routine (i.e., names of books read, drawings, shopping lists, cards, etc.).

b. parents and students complete a survey about literacy events in the home.

c. I speak to parents at parent-teacher conferences about what is read at home.

d. I ask parents to contact me to discuss any problems their children are having with reading and writing.

18. My parent-teacher communication tends to focus on
a. the progress of each child in my classroom.
b. the progress of each child with specific academic problems.
c. information about class trips, picture day, and other general information.
d. children misbehaving or having problems.

19. Face-to-face meetings with parents of children in my classroom occur
a. every day at arrival and dismissal time.
b. during specially designed workshops for parents of children in my class-room around academic topics.
c. during school sponsored Back to School Nights and Parent-Teacher Conferences.
d. when I am having problems with a child in my classroom.

20. You would find these regularly in my classroom:
a. family poems, family totems, family trees, family memoirs, stories, and so on, which include genres and text forms characteristic of family cultures.
b. parent story tellers reading and telling stories characteristic of their culture.
c. books characteristic of family culture sent in by parents to share with the class.
d. favorite family story books.

To score your survey: For multiple choice questions 1 through 16, give yourself 4 points for each a, 3 points for each b, 2 points for each c, and 1 point for each d you chose. For multiple choice questions 17 though 20, give yourself a point for each circled response.

If your score was 20–35: Chances are you find it difficult to work with the parents of children in your classroom. You might find them intimidating or too controlling. It would be helpful to find ways you can be comfort-able with the parents to begin a partnership. You can establish a regular routine of read-alouds and invite a parent to come and read to the class. Students in your class can work with their parents on a family memoir. The piece can have two voices, the parents' and the students', giving two perspectives on a same family event. The student can share the memoir with the class.

If your score was 35–50: You probably value the parents of the children in your class but don't want them to be a controlling presence in your classroom. The suggestions you make to parents may help to support what you do in the classroom; however, you still feel it is your responsibility to control all activities. You might want to suggest books, activities, websites, and even community resources where parents and their children can work together around topics being studied in your classroom.

If your score was 50–65: You value parents and see them as partners in their children's academic success. You are pleased by the work they do at home that enhances classroom learning. You are ready to integrate home literacy events into your classroom. You might want to survey parents about the literacy events that occur at home and find ways to use them in your classroom. Parents are ready to hear how the work they are doing at home and at school meet standards set by the school, district, and state. Classroom workshops and after school events may work for you.

If your score was 65–80: Wow! You have established a strong parent-teacher partnership. You have worked side by side to establish classroom routines, curricula, and academic choices. The parents are aware of the standards and how they are implemented, how they are met at home and at school. It may be time for you to establish a school-wide program. Think of ways that literacy events can be shared with families across the grade or school. Biography Nights, Poetry Nights, Scrap Booking Nights, and other interests of your school community are excellent ways to help further connect school and home.

Suggested Resources

Endrizzi, C.K. (2008). *Becoming teammates: Teachers and families as literacy partners.* Urbana, IL: National Council of Teachers of English.

Henderson, A.T., Mapp, K.L., Johnson, V.R., & Davies, D. (2007). *Beyond the bake sale: The essential guide to family-school partnerships.* New York: The New Press.

Kyle, D.W., McIntrye, E., Miller, K.B., & Moore, G.H. (2006). *Bridging school & home through family nights: Ready-to-use plans for grades K-8.* Thousand Oaks, CA: Corwin Press.

Winston, L. (1997). *Keepsakes: Using family stories in elementary classrooms.* Portsmouth, NH: Heinemann.

References

Auerbach, E.R. (1995). Which way for family literacy: Intervention or empowerment? In Lesley Mandel Morrow (ed.), *Family literacy: Connections in schools and communities* (pp. 287–303). Newark, DE: International Reading Association, Inc.

Baker, L., Sonnenschein, S., Serpell, R., Scher, D., Ferandes-Fein, S., Munsterman, K., et al. (1996). Early literacy at home: Children's experiences and parents' perspectives. *The Reading Teacher, 50,* 70–72.

Briggs, N., Jalongo, M.R., & Brown, L. (1997). Working with families of young children: Our history and future goals. In Joan P. Isenberg & Mary Rench Jalongo (eds), *Major Trends and Issues in Early Childhood Education: Challenges, Controversies and Insights* (pp. 56–70). New York: Teachers College Press.

DeBruin-Parecki, A., & Krol-Sinclair, B. (Eds). (2003). *Family literacy: From theory to practice.* Newark, DE: International Reading Association.

Enz, B.J. (1995). Strategies for promoting parental support for emergent literacy. *The Reading Teacher, 46,* 168–170.

Epstein, J.L. (2001). *School, family, and community partnerships: Preparing educators and improving schools.* Boulder, CO: Westview Press.

Hoover-Dempsey, K., & Sandler, H. (1997). Why do parents become involved in their children's education? *Review of Educational Research, 67,* 3–42.

Isenberg, J.P., & Jalongo, M.R. (1997). *Creative expression and play in early childhood.* Englewood Cliffs, NJ: Prentice Hall.

Jensen, D.A. (2006). Using newsletters to create home-school connections. *The Reading Teacher, 60,* 186–190.

Lareau, A. (1989). Family-school relationships: A view from the classroom. *Educational Policy, 3*(3), 245–259.

Lazar, A., Broderick, P., Mastrilli, T., & Slostad, F. (1999). Educating teachers for parental involvement. *Contemporary Education, 70*(3), 5–10.

Martin, E.J., & Hagan-Burke, S. (2002). Establishing a home-school connection: Strengthening the partnership between families and school. *Preventing School Failure, 46,* 62–65.

McCarthey, S.J. (2002). Home-school connections: A review of the literature. *The Journal of Educational Research, 93,* 145–162.

Moll, L., Amanti, C., Neff, D., & Gonzalez, N. (1992). Funds of knowledge for teaching: Using a qualitative approach to connect homes and school. *Theory Into Practice, 31,* 132–141.

Morrow, L.M., Paratore, J., Gaber, D., Harrison, C., & Tracey, D. (1993). Family literacy: Perspectives and practices. *The Reading Teacher, 47,* 194–201.

Paratore, J., Homza, A., Krol-Sinclair, B., Lewis-Barrow, T., Melzi, G., Stergis, R., et al. (1995). Shifting boundaries in home and school responsibilities: The construction of home-based literacy portfolios by immigrant parents and their children. *Reading in the Teaching of English, 29,* 367–389.

Paratore, J.R. (2002). Home and school together: Helping beginning readers succeed. In Alan E. Farstrup & S. Jay Samuels (eds), *What Research Has to Say About Reading Instruction* (pp. 48–68). Newark, DE: International Reading Association.

Purcell-Gates, V. (1996). Stories, coupons and the TV Guide: Relationships between home literacy experiences and emergent literacy knowledge. *Reading Research Quarterly, 31,* 406–429.

Shartrand, A.M., Weiss, H.B., Kreider, H.M., & Lopez, M.E. (1997). *New skills for new schools: Preparing teachers in family involvement.* Retrieved January 6, 2009, from http://www.ed.gov/pubs/NewSkills/title.html.

Shockley, B., Michaelove, B., & Allen, J. (1995). *Engaging families: Connecting home & school literacy communities.* Portsmouth, NH: Heinemann.

Taylor, D., & Dorsey-Gaines, C. (1988). *Growing up literate: Learning from inner-city families.* Portsmouth, NH: Heinemann.

Tuten, J. (2007). "There's two sides to every story": How parents negotiate report card discourse. *Language Arts, 84*(4), 314–324.

Villegas, A.M., & Lucas, T. (2002). Preparing culturally responsive teachers: Rethinking the curriculum. *Journal of Teacher Education, 53*(1), 20–32.

Winston, L. (1997). *Keepsakes: Using family stories in elementary classrooms.* Portsmouth, NH: Heinemann.

9

Teaching under the Accountability Umbrella

While I hope to guide my students in developing a lifelong love of reading and writing, I know that more frequently I am helping my students reach certain state-mandated academic benchmarks by teaching them how to pass standardized tests. In the past, preparing my students for high-stakes tests often occupied a majority of the time in my classroom and allowed me little time to help them develop a love of reading and an ability to think critically…As a result of being indoctrinated into a system in which creative and critical thinking is an after thought, my students crave structured assignments with clear-cut right and wrong answers.

—(Samantha, elementary school teacher)

Samantha, a teacher in a modified setting in an urban elementary school, is clearly concerned with the big picture: teaching literacy strategies that will allow the children in her classroom to become confident, independent, engaged, and fluent readers and writers. She wants her students to think critically and creatively but believes that taking time away from her regular instructional program to teach to the test discourages what she is trying to develop. In fact, her students now want instruction that does not demand them to engage in critical and creative thought. Because of the demands of teaching to the test, Samantha is in her discomfort zone.

Challenge and Opportunity

The concerns Samantha has for developing students who are lifelong readers and writers is often in conflict with instructional practices aligned with high-stakes tests under the umbrella of teacher accountability. Although their initial teacher preparation programs have exposed teachers to theoretical and conceptual frameworks for reading instruction, new teachers often find it difficult to put their academic knowledge into practice when they feel they need to stop regular instruction and teach to the test. New teachers need time, support, and opportunities to translate their prior theoretical understandings into effective practice especially when adapting those practices to teach in an era of accountability. Too often what teachers are told, and even mandated, to implement inside their classrooms are reading curricula that are narrowly focused to tests of accountability. Teachers feel disconnected from what they believe to be best for their students in order to optimize their students' learning and the curriculum they are given that has been designed around high-stakes assessments. A dissonance exists between teaching to the test and how they want to teach often causing teachers to engage in practices that are counter to their own educational philosophies for the sake of raising test scores.

Opportunities exist for teachers to use data from high-stakes test to inform classroom instruction, not in a scripted or dumb-down version of a curriculum, but in a manner that allows them to integrate and adapt a wide range of instructional strategies into data-driven lessons. Teachers need assistance in navigating the accountability issues and the curriculum to teach to the content of the test effectively without losing their curricula and instructional decision making freedoms. Too often teaching to standards and the standardization of curriculum is confused but it is not the same. The standards are the goals the state, district, or school aspires to obtain for their students. Standardization is the uniformity in which the standards are taught. The dilemma causes tension and anxiety for schools trying to promote student learning and achievement.

How do teachers know they are teaching children the strategies they need for future learning and not just to pass this test? Why do teachers feel the need to focus on test prep? Many teachers feel discomfort with the increasing demands of high-stakes assessments. One of the biggest challenges many teachers face is balancing what the administration expects of them in terms of accountability and what they know the children need beyond test preparation.

Story from the Field

Barry is beginning his first year of teaching. He was interviewed and offered a position in a third grade classroom at the school in his neighborhood, just a few blocks away from his apartment. He loves this part of the city, the diversity, the many shops along the avenue, its proximity to public transportation, the young singles and families that occupy many of the neighboring buildings. He could not have been happier with the way things worked out since graduating a few months ago.

Before school began, Barry spent a few days getting his room ready for the children by leveling and shelving books left by the retiring classroom teaching and from his own collection. He decorated the bulletin boards, made literacy centers around the room, and made name plates for each of the children. He was ready to begin his first year, confident of his undergraduate preparation, his success at student teaching, and his passion for learning and teaching. However, as the school year began, Barry felt overwhelmed. His literacy centers sat idle. The pile of great children's books that he was excited about started to collect dust. He wondered if he would ever get to do book talks. He found himself sticking to the teachers' guides, struggling with classroom management, and keeping up with workbook pages and grading. He kept moving the pile of children's books from one side of his desk to the other as he prepared for curriculum night and again at parent-teacher conference time.

By November he began implementing the literacy practices he learned in his college courses. He used the literacy centers, students were in book clubs reading children's literature, he held regular reading and writing conferences, and even started a classroom newsletter with student input. Barry was starting to focus on his students' learning and adapting his instruction to their needs. Then...POW!

Right after winter recess at the monthly faculty meeting, the principal told the staff they needed to get the children ready for the state tests. All regular instruction needed to be minimized. All teachers would receive test prep materials for their class, which they were to begin using immediately. The principal reminded the instructional staff that their school was on probation. He felt that the new scripted test prep materials would better ensure higher scores by students on the test.

Barry left the meeting in high anxiety. Third grade was an important testing year in his state. He had just begun to feel in control of the third grade curriculum and his ability to adapt the materials and his instructional strategies to meet the needs of his students. Frustrated at having to abandon all instructions that he had worked so hard to get in place, Barry went home and started to read all the test prep material given to him at the meeting and sighed. How was he ever going to get his students ready in time?

Research and Reflection

No Child Left Behind Act (NCLB) of 2001 has refocused our attention on test scores, accountability, and their uses in planning and implementing instruction. Since test scores and adequate yearly progress determine the extent of federal and state involvement, school administrators are very concerned with the test results. The use of external accountability mechanisms, such as high-stakes tests, has transformed educational practices in an effort to make both teachers and students accountable for their performance (Diamond & Spillane, 2004). Using test scores as the criterion for effective teaching and student learning has put undue pressures on teachers. As a result, many reading specialists and teachers are experiencing the tension between developing assessments and methods of instruction responsive to individual children's needs and implementing those they are mandated to use. Often the mandated curricula are in the form of commercially instructional packages or may be narrowly focused on test preparation; and often what reading specialists believe they should be doing, believing, and thinking conflicts with their existing practices and experience.

Tension in Accountability

Many teachers report abandoning what they know to be the best ways to teach reading in exchange for test preparation curriculum designed to raise test scores (Hollingworth, 2007; Urdan & Paris, 1994). They have reported that accountability has narrowed the curriculum, adversely affecting what they teach, the quality of instruction, and instructional time (Anderson, 2009). In a survey of teachers, Urdan and Paris (1994) found that four out of five teachers surveyed took time away from their regular classroom instruction to teach to the test. Teachers feel the need to emphasize specific information that will be on the test while sacrificing material involved in critical thinking, problem solving, and student interest (Anderson, 2009; Farstrup, 2006; Hollingworth, 2007; Wills & Sandholtz, 2009).

Being judged by their students' test scores have left teachers feeling humiliated and many are losing their confidence in what they believed was best for their students. Altwerger and colleagues (2004) found that schools preferred teachers who unquestioningly followed teacher manuals and curriculum guides over those who were creative, innovative thinkers, and decision makers. Similarly, there is a disconnect between specific scientifically based reading assessments, required for external accountability, and the information teachers need from assessments to plan appropriate daily instruction for their students (Invernizzi et al., 2005). "Teachers have lost flexibility in choosing appropriate assessments or developing instructional approaches that fit the strengths and needs of an individual child. Often they must use the same packaged program for every child" (Altwerger et al., 2004, p. 127).

According to Yatvin (2008), NCLB mandates focus not on student learning but on student achievement measured in number. It has supported a scripted program, a curriculum designed for rote, step-by-step learning when years of research on child learning has taught us children do not learn in that way. "Although scripted programs for reading have now returned stronger than ever to help with standardized testing, they have taken on a mechanical quality that truly is taking the life out of teaching and learning" (Wepner, 2006, p. 141). Some teachers report having to teach to the test all year, while others report having to implement test prep materials for a predetermined amount of time before the test is given. These overly directed materials fail to acknowledge the teacher as the qualified decision maker of what, when, and how reading should be taught thus reducing the control and autonomy over their classroom practices (Wills & Sandholtz, 2009; Yatvin, 2008). This dichotomy has caused teachers to doubt themselves and what they know about student learning, teaching demands, contextualized learning, and the personalized treatment of students. Teachers are often left feeling frustrated because of having to teach to a test rather than teaching for student success (Vayo, 2008).

An overemphasis on teaching to the portions of the reading curriculum that are testable on high-stakes tests has not only narrowed the curriculum but goes against the way teachers know is the best way to teach (Hollingworth, 2007) causing them to lose their confidence. Not given the flexibility at their schools, many teachers are uncomfortable planning, not just implementing, an instructional program for their children that prepares them for the test and for authentic reading and writing tasks. Teachers need the opportunity to refocus and to reflect on what should inform their instructional practice.

Teachers and school administrators have been held accountable for student achievement and base their success (the teachers', the students', and the administrators') on a test score. According to Hoffman (2004) a popular belief is that student test scores can be raised by giving harder and harder tests and holding everyone accountable for raising scores. According to Wilson et al. (2005) numbers are driving the curriculum but do not provide a complete picture of children's literacy capabilities. They go on to state, "While numbers certainly help in making broad program and policy decisions, demonstrating accountability by only using assessments that distill children's reading to a number obscures understanding of the differences between and amongst children as individual readers" (p. 622).

Understanding Accountability and Assessment

High-stakes testing is being used to promote higher educational standards. The high-stakes tests given to measure student achievement are aligned to standards. The standards also determine the tests and often influence the materials used in the classroom. One of the aims of accountability is to ensure that all students receive high quality instruction and reach a certain level of competence (Diamond & Spillane, 2004). They are also used to assess teacher effectiveness, compare schools within a district, and occasionally to evaluate new curricula (Urdan & Paris, 1994). Some agree that standards and the methods of accountability lead to improved student learning and reduce the inequalities in our educational system (Murnane, 2000) while others feel they are harmful to teachers and students (Ohanian, 1999).

A response to standards and their accompanying assessments by administrators has been to centralize or standardize a curriculum and prescribed instructional strategies as a simple and direct means of increasing student performance (Wills & Sandholtz, 2009). The standardization, in response to uneven student achievement and teacher quality in some schools, has dictated how teachers should teach (Hirsch, 2008; Hollingworth, 2007). Just as teachers need to assess the assessments, they need to look carefully at the data from tests of accountability and use and supplement them with those that will yield the data they need to inform their practice (Nelms, 2004). For the novice teacher there is a tension that forms when their teacher preparation programs teach the use of performance-based assessments and the schools primarily use standardized tests to measure student learning (Crumpler and Spycher, 2006; Darling-Hammond, 1997). However, scientifically based reading instruction has been equated with narrowly focused, scripted reading instruction (Farstrup, 2006) and Darling-Hammond (1997) warns, "what can be easily scripted, sequenced, and tested in a standardized fashion tends to represent only the most trivial aspects of the underlying knowledge sought" (p. 51). The key for many teachers is to document the literacy development of students and taking the time to organize the data effectively to inform instruction (Cobb, 2003; Johnston, 2003; Mokhtari et. al., 2007).

Teachers need to adopt and implement a multidimensional approach to assessment and to value the information each high-stakes test yields in order to integrate the findings into a carefully constructed profile of the children's strengths and vulnerabilities. Formal and informal assessments do not have to be at odds with each other. "It allows us to reflect an image of competence and agency by revealing to learners what they are going well, and it balances the high-stakes testing central focus on attending to what less capable students are not doing well" (Johnston, 2003, p. 91). A wonderful suggestion from Toll and her colleagues (2004) was not to seek a solution to the dilemma, to the conflicts and tensions existing between what is mandated and what is believed to be best for students, but rather to make the conflicts visible, create spaces in instruction for complexity and competing desires. They go on to suggest exploring the power available to the teacher and looking for dichotomous thinking and seizing the opportunity to discover any overlap.

Using the Data from Accountability and Assessment

Accountability and assessment need to inform practice that needs to inform further assessment. Cobb (2003) reminds of the reciprocal and integrative nature of effective teaching and learning in the relationship of curriculum, assessment, and instruction. Accountability must be part of the reciprocal and integrative nature as well. When studying how schools in Chicago used the data from high-stakes tests, Diamond and Spillane (2009) found a different response in low-performance schools than in those schools deemed as high-performing. The failing or low-performing school did not use the test data to inform instructional changes. High-performing schools used their test data to define specific instructional needs. They also used the data to make school level instructional decisions. Conversations around data to improve teaching and learning are not only productive, but acknowledge the importance of community in responding to the data. Many principals of high-performing schools have rejected implementing a standardized curriculum in response to test data. They see increased student performance as a joint effort. They rely on teacher expertise and judgment regarding student learning and support teacher control over the curriculum (Wills & Sandholtz, 2009).

For data to be useful to teachers and administrators alike, familiarity with the validity and reliability of any test as well as information about population sample is essential (Invernizzi et al., 2005). They can also establish how closely the test is tied to their current instructional program. This information is needed to plan instruction and instructional programs.

For data to effectively inform teaching, teachers need to make distinctions among the results they collect from a variety of assessment tools and high-stakes tests. The numerical score on a test does not demonstrate how a student engages in real world literacy demands (Johnston, 2003). "Focusing on numerical scores (e.g., a couple of scores from similar types of tests) increases the probability that valuable information about children and their reading instruction will be lost, especially when dealing with a complex act like reading" (Wilson et al., 2005, p. 624). Yet knowing what the test measures and the numerical score indicate will aid teachers in determining what they should teach to help students become proficient readers. Standards can frame and guide teachers' work without constraining it. The results from standardized tests, running records, informal reading inventories, writing samples, and the many other data collected on a child need to be viewed both separately and collectively. Similar to looking at a collection of photographs individually and together to reconstruct an event, the teacher must look at the data yielded from each assessment tool. Teachers must develop a clear understanding of the assessment, the curriculum standards and then carefully integrate these into instructional activities that demonstrate best practice. Increased knowledge of the standards and of the tests allow teachers to see instructional opportunities through a new lens. It assures teachers they are teaching the content of the test without having to teach to the test.

There is no evidence that the amount of time spent on test preparation improves test scores or students' reading proficiency (Allington & Cunningham, 2007). However, according to Anderson (2009), teaching a curriculum that is aligned with state standards and using test data as feedback has a positive effect on test performance. He suggests that teachers are better off in aligning their curricula to standards and use test data to make instructional decisions rather than spend time teaching to the test.

Teaching for Content and Not Just the Test

There is a dissonance between teaching to the test and how teachers want to teach. Teachers feel their instructional program is usurped by the need to teach to the test. By looking at test data, test content, and what is expected of students teachers begin to look at the bigger picture and where those fit in their regular curriculum. Teachers need to decide where to begin to make a difference, to prioritize instructional goals based on data of each individual student and their class collectively.

Teachers need to know how to integrate the uses of high-stakes testing to inform their practice. Some teachers have elected to teach testing as a separate genre (Hornof, 2008; Santman, 2002). Hornoff (2008) uses the test prep material and has her class study the special features and vocabulary unique to the test her students will be taking. Santman (2002) used test genre study within a workshop model already established in her classroom for reading. Students studied the materials designed to help them negotiate the test. In groups they studied how the materials were organized, the types of questions asked, and how to select an answer. In both instances, teachers were preparing their students to be informed consumers of tests by making them aware of the format and helping them to be good test takers.

Teachers should develop a sense of informed, professional decision making instead of relying on instincts (Baines & Farrell, 2003; Clandinin & Connelly, 1992) when designing instruction in response to accountability and classroom expectations for student learning. Not only should teachers be implementing instruction for students to acquire the necessary literacy strategies they need now, but to also teach to prevent future literacy struggles. As Johnston (2003) stated, the focus on high-stakes testing becomes balanced when teachers recognize patterns in their instruction while adapting their teaching to what children can do independently, almost do, or fail to understand. Through a process of pedagogical reasoning teachers scrutinize material to be covered, decide what is worth teaching, organizing and segmenting the materials to make it more suitable for teaching (Wills & Sandholtz, 2009).

Good teachers use a variety of assessments to make sound instructional decisions and to inform their teaching. "Professionalism demands thoughtful, grounded actions under complex and uncertain conditions that are nevertheless guided by, rooted in, and framed by professional standards" (Shulman, 1999, p. xiii). Effective literacy teachers welcome assessment and accountability programs that provide feedback on results and use it as a basis in their teaching (Farstrup, 2006) and make decisions and deliver instruction based on research-based practices that meet the needs of their individual learners (Crumpler and Spycher, 2006; Danielson, 1995). Certainly if we are to leave no child left behind, we need to look at each child's individual strengths and vulnerabilities as a learner in relationship to the standards within the classroom, school, district, state, and tests of accountability.

Often only minor adjustments can be made to current instruction if teachers take the time to familiarize themselves with the test. Hollingworth (2007) suggests conducting an alignment study between the curriculum standards evident on the test with teachers' current lesson plans. This will help determine what is being taught and what is being tested.

Teachers can integrate authentic reading instruction into test passages on test prep materials. By tailoring and adapting materials to students, the teacher is making the content ideas accessible to all students. The type of reading required on most standardized tests is unlike the real reading required of students in classrooms. To narrow the authentic reading tasks by imposing worksheets or multiple choice activities would send the wrong message to students about the reasons we read.

"Teachers must also be free to use material that allows them to connect what must be taught with what students can understand" (Darling-Hammond, 1997, p. 232). The freedom to plan a curriculum for a specific group of students, rather than having a curriculum guide determine the materials to be used and the lessons to be implemented, makes a teacher's time with the children more effective. The flexibility in deciding how to teach allows teachers room to adapt to the particular content of the test and the students being taught with room for exploration. This way the teachable moment is not lost to externally determined activities.

A good teacher, like a good actor, can put life into a script and Commeyras (2007) proposes asking us, "How does the teacher modify the lesson plan or script to provide differentiated instruction for students? Does the degree of specificity in a planned or scripted lesson need to differ depending on the instructional goals?" (p. 407). These questions are critical in working through the tensions of teaching under the accountability umbrella. Wepner (2006) suggests teachers must understand how to use a wide range of instructional strategies to address standards that are measured through testing. She believes teachers should construct lesson plans that reflect curriculum goals and include test prep strategies and state standards. Understanding the data and knowing what the tests measure are important for teachers to keep in mind when implementing classroom instruction. In this way the teacher can judge the level of student understanding and make space for student comments and questions.

Teachers and children thrive when they are given an opportunity to make real choices. Ivey (1999) found that students want to be and can become good readers when they are motivated with real purposes for reading, provided with a wide variety of materials of their interests, and the opportunity to collaborate with their peers. Reading that interests students help them develop a lifelong love of reading. Knapp (1995) found high-achieving schools allocated blocks of time for literacy teaching and learning and children spent more time actually reading than did children in lower-achieving schools. By making time for children to engage in authentic literacy activities, the teacher is effectively teaching to the test by teaching content.

The test is not the problem. In fact the scores from tests can be helpful to teachers. "Educators should hold their ground as professionals and refuse to compromise their teaching practice in the name of higher scores" (Hollingworth, 2007, p. 342).

Revisiting the Field

Barry began by reading through the packet given to him by the principal. It was a compilation of reading passages, grammar worksheets, vocabulary sheets, and word study skill sheets. There were also instructions about the writing portion of the test. Included in the packet were mathematics worksheets as well. He looked through the packet carefully and noticed that other subject areas were missing. There were no prep materials for science or social studies. What was this test his third graders would be taking next month? He was unfamiliar with the test or the format. Barry didn't want to call a colleague and appear to be incompetent. He decided to go online and google the name of the test just out of curiosity.

A number of sites popped up from his search. He chose to go to his state's education site for more information. Barry learned that the test was designed to give schools information about each student, what each student should know and be able to do by the end of each grade. He read that his grade, third, would only be tested in the areas of Language Arts Literacy and Mathematics, which explained the void of other subject areas in his test prep packet. He also found that the test relied on two types of questions for students: multiple-choice to uncover the student's ability to work with and analyze text, and open-ended questions requiring a short or long written response about a picture and a poem. Hmmm, Barry had planned to start a unit of poetry next month. Perhaps he should begin earlier?

Barry decided to read on. He learned that the test was field tested for two years and the state had only been using the test since the enactment of NCLB. The construction of the test began with a formal review of the Core Curriculum Content Standards and after the test items were developed, each item was aligned with the Core Curriculum Content Standards to ensure validity. He also read about the test's reliability, how students with disabilities and English Language Learners (ELLs) would not be excluded from the test but may be accommodated if they had an Individual Educational Profile (IEP). He also read how the results would be reported to the parents, the school, and district.

Feeling much better informed about the test his students would be taking in about six weeks, Barry took another look at the prep packet. If, in fact, the test was designed to measure how well the student is doing to meet the language arts standards for his grade, why not take a look at the Standards as well? Barry called Suzanna, a colleague and also a third grade teacher, and asked if they could meet. He told her the plan he had of analyzing the test to see how it aligned with their third grade standards to discover what they were already doing in the classroom to prepare their students for the test and where they might need to infuse some extra time for the students to be more proficient when testing day arrived.

Barry, although still anxious, felt he had a much better grip on the test, what he needed to do for his third graders, and how he needed to proceed with test preparation. Maybe he would even get to the pile of children's literature sitting quietly on the shelf behind his desk.

Strategies for Teaching with High-Stakes Testing

Like Barry, many teachers find themselves at the edge of their comfort zone when faced with preparing students for high-stakes tests. When handed a test packet and told to abandon their regular instructional programs, teachers become anxious and feel they have lost their professional voice in their classrooms. The unspoken message is that what they have been doing is not good enough for preparing students for the test. They are asked to change what they are doing for the sake of improving student scores, to narrow their curricular focus to only those testable items.

Vicky, a fourth grade teacher remarking about accountability, said,

> I think the main problem here is not so much that teachers are teaching "to the test" but the fact that so many teachers feel that they are wasting time teaching to a test that is in no way beneficial. I wish that there was a way to make these standardized tests more in sync with curriculum and/or if the test content was more meaningful and actually tested what is important to know in the form of a reliable and valid test.

Vicky speaks for many teachers. She is not aware of how the test is, or perhaps is not, aligned with her state's standards. She is also not aware of how the test has established validity or reliability.

Becoming familiar with the test, the reasons behind it, how it is constructed, what it measures, and how results are reported eases some of the anxiety teachers have about it. Instead of focusing on getting ready for the test and what the results might mean for an individual teacher or school, teachers need a basic understanding of the test itself before moving forward. Some tests are designed to measure students' competence against state standards by grade level, other tests to measure student achievement based on prior performance. It can differ state to state.

The Tool for Practice *What About the Test?* found at the end of the chapter will help teachers discover what they know about the test their students are required to take as well as help to identify the challenges or strengths the students already possess. It will also guide them in making discoveries about the test itself, what they need to know to feel more comfortable about test design, and what it might mean for the teacher and the students.

Toni made the point after reflecting on a discussion of accountability measures and results with the teachers at her school by stating,

> I think that there is a way to use data from standardized tests to reflect upon the curriculum and how it is meeting state standards and the effectiveness of teacher instruction. But looking at the test data requires specific skills and

teachers must be given insight into how the data was sorted and aggregated. My concern about too much emphasis being placed on high-stakes testing is it assumes that the curriculum is deficient in some way.

Teachers can study the test and the curriculum to ensure that the curriculum content they deliver is not deficient in relationship to the test.

The high-stakes tests given in any state are aligned to the state standards. Just as Barry found the test items given to his students were aligned to the Core Curriculum Content Standards in his state, teachers need to know how test items on their required tests are aligned to their state's standards as well. By studying previous years' tests and comparing items to the state's standards teachers can discover the test's emphasis. Does the test focus on vocabulary? Does it require students to recognize an answer or produce one of their own?

Caryn, a second grade teacher, in talking about test preparation told us,

> One of the biggest challenges that I faced was balancing what administrators expected of me and what I knew the children needed from me. I still grapple with this today. So, if I decide to do something that may not be included in the curriculum, I prepare my reasons for doing it and how it aligns with the State Standards. If anyone happens to walk in, I have a rationale prepared.

Grade level teachers and their literacy coaches can study the focus and components of the test, how state standards align to test components, and how instructional materials and lessons already present in their classrooms prepare students to meet the standards and ultimately succeed on the test. Locating gaps between components of the test, standards, and classroom instruction could explain low performance. Finding the gaps will assist teachers in the development of future lessons to ensure they are teaching the content of what will be on the test.

The Tool for Practice *How Instruction Aligns with Mandates* at the end of this chapter is designed for this purpose. By determining and reviewing the connections among tests, standards, and classroom instruction, teachers often discover they are teaching the content of the tests and meeting standards. The tool also allows teachers to determine where additional lessons are needed and can then design and implement supplemental lessons to ensure student success. However, alignment studies require team effort. Determining which items contained in the standards are also addressed or not addressed in classroom instruction is labor intensive and best done in groups.

Teachers collect data about their students' literacy progress throughout the year. The classroom assessment tools demanded by the school and currently present in classrooms, such as running records, retellings, and other informal assessment, yield data that teachers use to drive instruction. By collecting and analyzing data on each child in your class, you can determine patterns of learning and identify specific areas of need for the whole class, a small group of children, or an individual child, as discussed in chapters six and seven. By tracking a child's progress over time, adapting instruction, and giving feedback to the student his or her strengths and weaknesses can be identified. They can serve as a predictor of how well the student will do on the test and can assist teachers in targeting instruction.

Rickie, a teacher in an urban public school, stated,

> Assessment is absolutely essential when it comes to teaching children. First of all we need to know where they are in terms of their capabilities, so that we can push them towards where they need to be. That is why authentic assessments are so important. They cater to the individual as opposed to the standardized one size fits all approach. We know that it doesn't truly reveal the strengths and weaknesses as authentic assessments do.

Rickie fails to see the how feedback on the authentic or informal assessments she uses with her class can be of benefit when getting her students ready for the test.

The Tool for Practice *Literacy Goals and Classroom Assessments* helps you organize the data around test items and standards for each child in your classroom. By keeping track of students' progress toward areas being tested, teachers are in a stronger position to adjust the regular instruction rather than abandon the curriculum to teach to the test.

Some teachers teach testing as a separate genre taking time out of their instructional program to do so. Tom said, "I learned that it's important to teach test prep as its own genre. I found this idea to be helpful because it allowed me and my students to see that it's just one part of literacy, just like poetry, non-fiction and fiction. I think it also helped take some pressure off of them." By using test prep materials and infusing them with authentic reading instruction methods, teachers do not feel that they are totally abandoning their instructional program. By determining which reading strategies could be taught with test prep materials, teachers are creating opportunities for their students to develop abilities.

The tools at the end of this chapter are designed to empower teachers with the knowledge they need in a testing environment by looking at the tests, how test items align with standards and curriculum, and tracking student progress. Teaching under the accountability umbrella should not isolate or undermine teachers' confidence or their professional prerogatives. Accountability doesn't improve teaching. Only teachers can improve their teaching. It is time for teachers to take charge of the quality of instruction knowing that teacher expertise is the most important factor in student achievement.

Tool for Practice: *What About the Test?*

How much do you know about the standardized test(s) given at your school? Use the questions given in table 9.1 for discussion with a colleague or use

Table 9.1 What about the test? Developing standardized test knowledge

Test information	Answers	Is this a strength or challenge for my students?	How can or how is it addressed in my classroom?
Name of the test			
What is the test designed to do? What does it measure?			
What format does the test take? Is it all multiple-choice? Short answer? etc.			
What subject areas does it test? Are there subcategories of the test within each subject area?			
How do the test items align to the curriculum standards?			
Have validity and reliability been established?			
Who is required to take the test? Are special populations required or exempt from the test?			
How are the results reported? To whom are they reported?			

them to discover what you need to know about the test(s). Then decide if the elements are a challenge for your students or if they bring a strength to it. How is the element addressed (or not addressed) in your present classroom practice?

Tool for Practice: *How Instruction Aligns with Mandates*

This tool (table 9.2) is designed to help you identify the components of the standards, the focus of the high-stakes test your students will be required to take, and the instruction in your classroom that aligns to both test focus and standards. This is done most successfully with other teachers on your grade and with the help of the literacy coach. This collaborative effort opens up opportunity to develop a strong understanding of the test, standards, and the curriculum. The knowledge gained from this self-study can empower teachers because it acknowledges their instruction as important and allows for a wider view of their classroom practice in a climate that has typically narrowed the curriculum to subjects that can be tested.

In the first column record one area of literacy on which the test will focus. For example, the test given in your school, district, or state may focus heavily on Drawing an Inference as part of the comprehension portion of

Table 9.2   How instruction aligns with mandates

Test focus area	State standard	Instruction present meeting standard	Supplemental lesson idea

the test. In the second column record the standard or standards that align to that test focus. You might find that the standard is already present in a variety of curriculum areas such as social studies and science as well as for reading.

The third column of the tool allows you to go through your curriculum to determine where and how the standard, and ultimately the test items, are met by what you are presently doing in your classroom. What are the planned teaching and learning experiences that allow students the opportunity to acquire the strategies necessary to demonstrate adequate performance.

The fourth column allows you to note any gaps needed to be filled by supplemental lessons. When planning, think across curriculum or subject areas where these might logically and naturally be developed.

Tool for Practice: *Literacy Goals and Classroom Assessments*

There are a few parts to this tool (table 9.3) and you will need one table for each of the children in your class or for each child you see in a specialized setting. By completing the table early in the school year, you are able to see how your classroom assessments can measure probable success on the high-stakes test used in your school and meet standards. In this way you are assured that your instruction is already meeting standards and you will not have to "test to the test." The table will also alert you to the time when specific goals are being addressed in the year, giving you the opportunity to observe your students and determine how each is meeting the goals. Another benefit of the table is its ability to track students' progress toward each goal so you can identify early on which students own them and which students might be struggling.

In the first column list the literacy goals each child is expected to own at the end of the school year. These goals can come from your curriculum guides, the program guides, listed as items to be tested on different formalized tests, or those you have for the children. You might find that the same expectation appears in different formats. You may or may not want to list where the expectation appeared on your table.

In the second column list the characteristics or strategies the children will need in order to be considered successful at reaching the goal. These should be listed individually.

In the third column list the classroom assessment you will use to determine if the child owns the characteristic or strategy. There should be multiple opportunities for students to demonstrate their learning. For example, you might be using retellings to determine a child's ability to sequence story events, which is not only a standard to be met but also found on the high-stakes test required in your school.

The final three columns allow you to record the development of these characteristics and strategies in each child: Not yet developed (N), Developing (D), Owned (O). You might want to use different labels depending upon your state's reporting methods. For example, in one state the results for each child are reported as below proficiency, meeting proficiency, and above proficiency.

Table 9.3   Literacy goals and classroom assessments

Name of Child:_____

Program expectations	Characteristic of meeting standard	Classroom assessment used to determine success	N	D	O

Suggested Resources

Barrentine, S.J., & Stokes, S.M. (2005). *Reading assessment: Principles and practices for elementary teachers* (2nd ed.). Newark, DE: International Reading Association.

Caldwell, J.S. (2002). *Reading assessment: A primer for teachers and tutors.* New York: Guilford Press.

Calkins, L., Montgomery, K., & Santman, D. (1998). *A teachers guide to standardized reading tests: Knowledge is power.* Portsmouth, NH: Heinemann.

Use a variety of Search Engines to access information about the test(s) used in your classroom.

References

Allington, R.L., & Cunningham, P.M. (2007). *Schools that work: Where all children read and write.* Boston: Ally and Bacon.

Altwerger, B., Arya, P., Jin, L., Jordan, N.L., Laster, B., Martens, P., et al. (2004). When research and mandates collide: The challenges and dilemmas of teacher education in the era of NCLB. *English Education, 36*(2), 119–133.

Anderson, L.W. (2009). Upper elementary grades bear the brunt of accountability. *Phi Delta Kappan, 90*(6), 413–418.

Baines, L., & Farrell, E.J. (2003). The Tao of instructional models. In. J. Flood, D. Lapp, J.R. Squire, & J.M. Jensen (eds), *Handbook of research on teaching the English language arts* (2nd ed.; pp. 74–86). Mahwah, NJ: Erlbaum.

Clandinin, D.J., & Connelly, F.M. (1992). Teacher as curriculum maker. In P.W. Jackson (ed.), *Handbook of research on curriculum: A project of the American Educational Research Association* (pp. 363–401). New York: Macmillan.

Cobb, C. (2003). Effective instruction begins with purposeful assessments. *The Reading Teacher, 57*(4), 386–388.

Commeyras, M. (2007). Scripted reading instruction? What's a teacher educator to do? *Phi Delta Kappan, 88*(5), 404–407.

Crumpler, T.P., & Spycher, E. (2006). Assessment has a dual purpose in teacher preparation programs. In S.D. Lenski, D.L. Grisham, and L.S. Wold (eds), *Literacy teacher preparation: Ten truths teacher educators need to know* (pp. 92–101). Newark, DE: International Reading Association.

Danielson, C. (1995). Whither standardized assessment? In A.L. Costa & B. Kallick (eds), *Assessment in the learning organization: Shifting the paradigm* (pp. 84–90). Alexandria, VA: Association for Supervision and Curriculum Development.

Darling-Hammond, L. (1997). *The right to learn: A blueprint for creating schools that work.* San Francisco, CA: Jossey-Bass.

Diamond, J.B., & Spillane, J.P. (2004). High-stakes accountability: Challenging or reproducing inequality? *Teachers College Record, 106*(6), 1145–1176.

Farstrup, A.E. (2006). NCLB, RF, HQT, SBR, AYP: ASAP? *Reading Today, 23*(5), 22.

Hirsch, E.D. Jr., (2008). Plugging the hole in state standards. *American Educator, 32*(1), 8–12.

Hoffman, J.V. (2004). Achieving the goal of a quality teacher of reading for every classroom: Divest, test or invest? *Reading Research Quarterly, 39*(1), 119–128.

Hollingworth, L. (2007). Five ways to prepare for standardized tests without sacrificing best practice. *The Reading Teacher, 61*(4), 339–342.

Hornof, M. (2008). Reading tests as a genre study. *The Reading Teacher, 62*(1), 69–73.

Invernizzi, M.A., Landrum, T.J., Howell, J.L., & Warley, H.P. (2005). Toward the peaceful coexistence of test developers, policymakers, and teachers in an era of accountability. *The Reading Teacher, 58*(7), 610–618.

Ivey, G. (1999). Reflections on struggling middle school readers. *Journal of Adolescent and Adult Literacy, 42*(5), 372–381.

Johnston, P. (2003). Assessment conversations. *The Reading Teacher, 57*(1), 90–92.

Knapp, M.S. (1995). *Teaching for meaning in high-poverty classrooms.* New York: Teachers College Press.

Mokhtari, K,. Rosemary, C.A., & Edwards, P.A. (2007). Making instructional decisions based on data: What, how, and why. *The Reading Teacher, 61*(4), 354–359.

Murnane, R. (2000). The case for standards. In D. Meier (ed.), *Will standards save public education?* (pp. 57–63). Boston: Beacon Press.

Nelms, B.F. (2004). On the front line: Preparing teachers with struggling schools in mind. *English Education, 36*(2), 153–167.

No Child Left Behind Act of 2001, Pub.L. No. 107–110, 115 Stat. 1425 (2002).

Ohanian, S. (1999). *One size fits few: The folly of educational standards.* Portsmouth, NH: Heinemann.

Santman, D. (2002). Teaching to the test?: Test preparation in the reading workshop. *Language Arts, 79*(3), 203–211.

Shulman, L.S. (1999). Foreward. In L. Darling-Hammond & G. Sykes (eds), *Teaching as the learning profession: Handbook of policy and practice* (pp. xi–xiv). San Francisco: Jossey-Bass.

Toll, C.A., Nierstheimer, S.L., Lenski, S.D., & Kolloff, P.B. (2004). Washing our students clean: Internal conflicts in response to preservice teachers' beliefs and practices. *Journal of Teacher Education, 55*(5), 164–176.

Urdan T.C., & Paris, S.G. (1994). Teachers perceptions of standardized achievement tests. *Educational Policy, 8*(2), 137–157.

Vayo, A. (2008). The tangled web of standardized test culture. *Thought and Action, 24*, 135–141.

Wepner, S.B. (2006). Testing gone amok: Leave no teacher candidate behind. *Teacher Education Quarterly, 33*(1) 135–149.

Wills, J.S., & Sandholtz, J.H. (2009). Constrained professionalism: Dilemmas of teaching in the face of test-based accountability. *Teachers College Record, 111*(4), 1065–1114.

Wilson, P., Martens, P., & Arya, P. (2005). Accountability for reading and readers: What the numbers don't tell. *The Reading Teacher, 58*(4), 622–631.

Yatvin, J. (2008). Where ignorant armies clash by night. *Council Chronicle: The National Council of Teachers of English, 17*(3), 25–31.

10

Pulling It All Together beyond the (dis)Comfort Zone

"What goals have you set for next year?" Amy mulled this over for days trying to figure out what she would write for her annual self-evaluation. Everything seemed to be going well. What would her next step be?

—(Amy, fifth grade teacher)

After three years teaching fifth grade, Amy feels comfortable with her literacy instruction in her classroom. She feels she has made a strong connection between theory and practice, her students are engaged and achieving, and she has earned the respect of her colleagues and parents. She wonders what steps she needs to take to continue to grow as a professional.

Challenge and Opportunity

The challenge of this question resides in the underlying risk of becoming complacent when what was once uncomfortable becomes a routine part of one's professional practice. Both challenge and opportunity arise from within. When the edge of your (dis)comfort zone becomes your comfort zone, then what? Do you leave teaching and move on to a new challenge? Or do you begin to help others? Do you privately ruminate about what could be better? Or do you volunteer to make a presentation about what works?

Throughout this book we addressed points of tension in teaching—the (dis)comfort zones that many teachers experience as they move toward increasing levels of skill, depth, and expertise in working with their students. We explored the (dis)comforts and opportunities in owning discomfort, describing rather than labeling students, and making conscious choices about words that impact student learning. We also examined the power of simply acting ourselves into a new way of thinking, rather than thinking ourselves into a new way of acting. In each of these zones of our teaching as well as the others presented in this book, we offered tools and strategies to discover yourself and work toward solving your own teaching dilemmas to take your teaching to new levels of opportunity and insight.

At some point, most of us have felt the way Amy does...you have overcome some professional hurdles and you wonder where to go next. Your teaching feels good, the students are learning, and you look around you for another challenge. Now is the time to expand your role as a teacher in new ways. It is time to slide open the classroom doors of your teaching (Silva et al., 2000).

Story from the Field

Julissa was in her fifth year of teaching and beginning to feel like she had "been there and done that." She loved her classroom and the students in her fourth grade, inner city classroom. Despite the tension and stress of the fourth grade testing, she felt that she had reached a peak. Her students were testing well, which was a measure of teaching quality that her principal valued. In addition, parents in the third and second grades were already taking the principal aside to request her as their child's teacher. She prided herself on developing good relationships with her students and their families. Her students also worked hard and she challenged them to achieve goals that they didn't think were possible. By the end of the year, all her students were reading, writing, and engaged in learning on level or better, and she was proud to send them along to fifth grade, confident that they were ready for the challenges of middle school. It hadn't been easy. All of these gains and outcomes had been the result of her hard work and persistence and had come at the price of sleepless nights and harried days. So, why couldn't she just enjoy it? Why did she have this little nagging urge to do something more...something else...something...she didn't really know what.

Research and Reflection

When you reach the end of your comfort zone and are wondering what comes next in your life as a teacher, leaving the classroom or becoming an administrator are not your only options. Teacher leadership has a long history but is recently becoming more accepted and supported as schools move toward distributed leadership models of management and reform.

But what is a teacher leader? Barth (2001) defines teacher leadership quite simply as "making happen what you believe in." In practice, then, it means taking your convictions, beliefs, experiences, and wisdom of practice and sharing them with colleagues in such a way as to extend beyond your classroom into the lives, practices, and policies of others (Hatch et al., 2005).

Lieberman and Miller (2004) assert that teacher leaders can take on three distinct roles in their work: that of advocates, of innovators, and of stewards. Advocates are those who keep their focus on students' learning and serve, in a way, as a school's conscience (Ackerman & Mackenzie, 2006). For example, Andy in chapter two really struggled with the literacy learning of his students and balancing school expectations with his students' learning. By examining his choice of words, Andy realized that he was "stuck" on telling, rather than teaching. At the point when Andy shifted his thinking to focus on the students, he became their advocate. And, as revealed by the changes in his student Max, this advocacy stance opened up new possibilities for Andy's teaching as well as his students' learning.

Innovators are those who are creative in their instructional practices and serve by making suggestions and acting as catalysts for others' to think about their work in new ways. As an example, Barry in chapter nine was "caught between a rock and a hard place" in wanting his students to succeed in taking the state-mandated tests and not wanting to teach to the test. Barry began by working alone in analyzing the test purpose and questions. As he dug deeper he studied the way the test was matched to the state curriculum standards. At that point, Barry invited other teachers into the process of analyzing the curriculum standards and matching them to their grade level curriculum. At this point these teachers began to innovate. As they better understood the expectations for their students, they planned for what they needed to do to prepare their students without teaching to the test.

What Barry did without formally organizing his colleagues was to initiate a professional study group. A professional study group can be defined broadly as, "a group comes together to learn something…a meaningful chance to build expertise" (Walpole & Beauchat, 2008, p. 3). A professional study group can take on a variety of formats depending upon the goals of its members. For example, they can decide to study state or district standards, student achievement data, curriculum materials, professional books, or methods of instruction (Walpole & Beauchat, 2008; Walpole & Blamey, 2008). A study group allows its members the opportunity to openly discuss their beliefs and practices as well as find support for their instructional and teaching contexts.

Stewards are those who focus their efforts more in the policy arena by serving to improve the teaching profession through continued professional growth. Through all the teacher examples presented in this book, not one of them took on this aspect of teacher leadership. Perhaps, then, this is an area most in need of exploration. For example, the teachers we presented worked to solve their practical dilemmas and each of them came to terms with their own discomfort zones. They were all advocates and innovators in one way or another, but making the leap into the role of a steward means taking on a wider audience than one's classroom and one's school. Stewards move in front of an audience of professionalism by claiming their voice in local, regional, and national decision making venues. They speak on behalf of teachers and students, and move the teaching profession forward by influencing decisions and policies with the power of their own words and the reality of their experiences of teaching and learning.

Although none of the examples in this book speak directly to the role of stewardship in teacher leadership, some of our teachers moved into that role unexpectedly. Shelley was asked to make a presentation to the other teachers to report the findings of her inquiry study in which she tried to understand how well the teachers in her high school knew their students. The study drew some conclusions about how teacher familiarity affected student learning, particularly for those students whom the teachers knew less well. Shelley ended up coauthoring a book chapter about the study and its findings. As another example, Amy found herself organizing a parent workshop based on the information she was sharing with parents in her classroom newsletter. The workshop was so successful that she was asked to "take it on the road" to another school across town when a teacher in her school told a friend at another school how great the workshop was. As a final example of stewardship, Denise agreed to join her professor at a national conference to talk about her graduate work as professional development and found, afterward, that she was invited to serve on a committee for the organization. Taking on roles of advocacy, innovation, or stewardship is not as complicated or as formalized as you might anticipate. Sometimes moving into one of these teacher leadership roles is as simple as being open to opportunities as they come your way.

But why is it important for teachers to step forward as leaders to do these things? Opponents to teacher leadership might point out that administrators and staff developers should be responsible for these expanded roles of advocate, innovator, or steward. However, teachers contribute some unique qualities and perspectives. Teachers are often the most stable force in a school or district that is experiencing leadership and/or administrative changes (Harris, 2005). As a stable influence, teachers maintain their focus on student learning and often know a school's internal structures and relationships in a way and at a level that "makes things happen" effectively (Silva et al., 2000). The benefits of teacher leadership also extend into the schools in ways that are unanticipated. For example, some schools report a drop in absenteeism, more motivated teachers, and a transformational potential in the school that was a catalyst for school development and improvement (Harris, 2005). Thus, it is important for teachers to step forward as leaders because of the benefits to the school, the students, and to each other.

But being a teacher leader does not mean that all the benefits go to everyone and everything other than oneself. Teachers who step up to become leaders experience new opportunities to make a difference, expand their influence, and take on greater responsibility (Phelps, 2008). Teacher leaders report that they grow in leadership skills and develop perspectives that are more organizational (Harris, 2005). They also report a growing sense of confidence and increased motivation about staying in their profession as well as becoming more effective as teachers themselves.

Silva et al. (2000) made five assertions about what teacher leaders do that speak to the importance of their potential influence and benefits to schools, students, and other teachers:

1. Teacher leaders navigate school structures.
2. Teacher leaders nurture relationships.
3. Teacher leaders model professional growth.
4. Teacher leaders help others with change.
5. Teacher leaders challenge the status quo by raising children's voices.

When teacher leaders navigate school structures, it is not a matter of closing the doors and conducting business as usual. For example, Linda in chapter seven was conflicted about the guidelines she was given for implementing guided reading lessons. Her focus was on following the steps in the process, not on what her students were learning. Linda didn't huddle up behind closed doors and worry her way into a solution; she reached out to a colleague and conferred as a professional. Linda retained the structure of the school guidelines for conducting guided reading sessions but she shifted her focus to the students and, by doing so, navigated the structure to make it more meaningful and effective for her students.

When teacher leaders nurture relationships, it is not a matter of remembering birthdays and sending "get well" cards to colleagues. Rosa's relationship with Oscar, in chapter six, is a good example of nurturing a relationship by balancing an appreciation of Oscar's strengths with a deep commitment to his progress and growth. This same dynamic of appreciation and commitment to growth is what teacher leaders do when they nurture relationships with colleagues, parents, and students. However, in Rosa's example she went beyond nurturing her own relationship with Oscar, she promoted Oscar's relationship with another student, Christian. Just like Rosa in her classroom, teacher leaders take these same practices and apply them to their schools and communities. They nurture relationships within and across boundaries to promote teaching and learning. Sometimes this can be as simple as setting up cross-age tutoring groups, or as complex as bringing together a study group of multi-grade teachers who are concerned about a common topic, such as the success of English Language Learners (ELLs).

When teacher leaders model professional growth, it is not a matter of "watch me, and then do what I do." For example, Emily in chapter one was challenged by independent reading and the practice of using Post-it notes to foster discussion in her Collaborative Team Teaching (CTT) classroom. By keeping her focus on the students and a teaching practice that was recommended but not working for her, Emily realized that student-initiated questioning was her real challenge. By exposing her own discomfort and reflection on practice, Emily led the way for her CTT classroom team to develop a questioning rubric and, as a result, the students became more engaged in their reading overall. In turn, Emily facilitated her own learning as well as the learning of her co-teachers. The focus was the students, but the benefit was revealed by modeling her professional growth that led others to become better, more student-focused teachers.

When teacher leaders help others with change, it doesn't mean becoming an administrative cheerleader. When Kara, in chapter four, was offered the opportunity to provide feedback to a colleague's videotaped teaching segment, she worried about what to say to Laura. Kara's commitment in viewing Laura's videotape five times ensured that the feedback would address significant strengths as well as substantive suggestions for improvement. Kara was a true teacher leader in helping Laura to change her practice. Rather than fitting a teacher into a template of practice, Kara started from what Laura was trying to do and helped her to accomplish her goals.

When teacher leaders challenge the status quo by raising children's voices, they don't shout. In contrast, like Sue and Heidi in chapter five, teacher leaders challenge themselves first. Sue was clearly uncomfortable as her graduate class moved into an area of inquiry that focused on a work of art and personal engagement with the artwork to deepen meaning. She resisted and challenged the process, but she came to an understanding of herself as a teacher who always followed the rules. Sue realized that when there were no rules, she shut down and she began to question how this related to her teaching. By examining her own learning, Sue asked how she could lead her students into self-discovery and empowerment, rather than rule following. Sue's introspection led to her giving her students a more powerful voice in her own classroom and, ultimately, in challenging the status quo of imposing a curriculum on students rather than opening up opportunities for them to discover themselves.

These five assertions can be five ways for you to think about expanding your role as a teacher into areas that are beyond your (dis)comfort zone. You can assist others by navigating the school's structure to get things done. You can nurture collegial and collaborative relationships among teachers, families, and the community by making connections, suggesting processes, and model for others how to bring people together for a common goal. You can be a model for others' professional growth by "sliding open the door" of your own classroom for others to observe what you do and how you learn (Silva et al., 2000). You can help others find their way as change comes to your school or as teachers institute change on their own. Finally, you can keep the children at the forefront of your thoughts, actions, and words by speaking for them in contexts where they are not included. Teacher leadership can be a way for you to continue to grow and develop as a teacher, such that your (dis)comfort zones do not disappear. They simply take on new shapes and meanings.

Revisiting the Field

Julissa's school was a large one with four floors built around a central courtyard with approximately four classrooms on each side. There were just under one thousand students in the K-5 school, which meant that there were six or seven classrooms at each grade level and more than forty-five teachers, including specialists. Sometimes she didn't even know the names of the teachers in the other grades.

It was about 7 p.m., late to leave the building even for her. But in selecting a more direct route out of the building, rather than passing through the office to her mailbox on the way out, Julissa walked down a hallway that she wasn't used to traveling. Curious about student work, she couldn't help walking slowly, pausing to look at bulletin boards of student work and glancing into the few doors that were open. After admiring a particularly playful display of student responses to literature, Julissa glanced into an open door where another teacher was sitting at her desk with her head in her hands. Usually she wouldn't have stopped or disturbed another teacher but she announced herself and said that she was admiring the work display in the hallway.

It turned out that the teacher was new to teaching. The work display was one of her good ideas but she felt that she was drowning in a sea of expectations and it was taking a toll on her health and happiness. On the day when Julissa walked in, Amy Wong was experiencing a blow to her ego. Her students had been acting up, a parent was upset with a homework issue, her plans for the day were interrupted by a fire drill and it took a whole class period to get the students refocused. It seemed like she was explaining again and again the same concepts she had explained last week and the week before. Weren't they learning anything? What was she doing wrong?

Thus began an enduring relationship of sharing and conversation. For Julissa the relationship refocused her energies on learning from students as she worked with the new teacher on understanding the issues of her classroom. Together they planned, which helped Julissa's own planning. Together they questioned, which helped Julissa discover new questions. They worked together, created cross-age and cross-classroom partnerships in reading and writing, and began to do things that neither of them might have imagined on their own.

Strategies and Tools to Venture into Teacher Leadership

Teacher leadership can be fostered through a variety of roles and opportunities. For example, Julissa began to "coach" new or novice teachers. The "coach" role can be formally defined, such as the school's literacy coach or the professional development coach for third and fourth grades, or informally pursued, such as observing a child at a colleague's invitation because s/he is at a loss for how to get past a barrier to learning. Working in this way with other teachers is a form of teacher leadership.

The *Curricular Support Checklist* at the end of the chapter is a useful tool to explore the ways that you can support your colleagues in their classrooms. Think of this list as possible conversation starters when a fellow teacher is expressing concerns or frustrations, such as "You know what? Maybe I could help you figure that out...would you like to _____" (fill in the blank with one of the checklist items)?

Another possible way to take on a greater leadership role as a teacher is to lead an inquiry group study of an important question or topic of concern to your grade level or school. For example, if your grade level team of teachers consistently discusses how difficult it is to reach their students who are ELLs, then perhaps it is time to organize a small group to study the issues in a focused and well-organized way. Inquiry study groups or professional learning communities are growing initiatives in school-centered professional development. Taking a leadership role in these initiatives is an opportunity to continue to grow and learn as a teacher leader. A number of good resources for inquiry group work can be found online by simply typing "Teacher Inquiry" into a search engine such as Google. However, at the end of the chapter you will find *Eight Steps to Getting Started on an Inquiry Project* to get you started.

Teachers at one early elementary school were concerned that their students were not achieving as well on the state third grade English Language Arts test as they had hoped. With a large and growing population of ELLs, they decided to focus their attention on vocabulary development. With the assistance of their literacy coach and principal, they created a reading list of recent research on vocabulary and ELLs. Each Monday afternoon, after school, they met to discuss the readings and to make connections with their classroom experiences. As they became more informed, they decided to move their reading into action. Research suggested to them that they needed to strategically teach more specific words. They initiated a pilot program using science trade books in the first grade. Teachers carefully documented their instructional practices, observed each other, and observed and assessed children. Through their continued and collaborative discussions, they analyzed their data and found that their innovations were having an impact on the children's vocabulary growth.

As a third, and final, strategy for moving into teacher leadership there are opportunities to share what you know with other teachers. Teachers love to hear what other teachers have developed to solve problems in their own classrooms. Although it may be difficult in one's own school (it is hard to be a prophet in your own land), there are many, many opportunities to get involved in professional organizations to present and discuss your practice with other teaching professionals. For example, did you design an interesting curriculum unit that integrated science, social studies, and literacy? Then share it at one of the state- or national-level conferences of the reading, social studies, or science education associations (a list of content area professional associations is provided in the section on further resources at the end of this chapter). Presenting your work at the state level or nationally is not nearly as intimidating as it may sound. These professional associations are eager to include teacher-friendly and teacher-useful presentations in their conference programs. To get started, use the *Presentation Checklist* provided at the end of the chapter as a guide to focus and present your good work.

Throughout this book we focused on moving through and beyond your (dis)comfort zone toward a greater depth of understanding and effectiveness in your teaching. This chapter attempted to challenge you to move through and beyond the (dis)comfort zone of remaining within the safety of one's own classroom to take a greater leadership role within your school and the professional community beyond your school. Are you ready to be a teacher leader? Take the tongue-in-cheek survey for self-study quiz at the end of the chapter to test your own readiness and then build on your strengths as you move into the ever widening circles of professional practice.

Tool for Practice: *Curricular Support Checklist*

As a teacher leader you can offer help or invite a colleague to help you:

- Organize and set up the classroom library "Come on, we can do it together. Many hands make quick work."
- Create shared assessments. "I have that problem too, let's work on an assessment together."
- Analyze results from the shared assessments. "Help me figure this out."
- Plan instruction based on shared assessments. "Now what can we do?"
- Schedule instruction. "What about if we try it this way?"
- Match readers with texts. "Look at what I found. It really helps."
- Share reading and writing lessons. "Come watch me teach and tell me what you think."
- Set up the writing workshop. "Let's work this through together, I can use the support too."
- Work on lessons to write in a specific genre (poetry, nonfiction, fiction, memoir, etc.). "I want to explore memoirs, any ideas?"
- Create Units of Study with a specific genre. "Let's do a unit together."
- Set up the reading workshop. "Let's work this through together, I can use the support too."
- Design mini-lessons about comprehension strategies, word study, fluency, decoding, and so on. "My mini-lessons always take longer than I intend, too. Let's see if we can work out a way to keep them as mini-lessons rather than lesson-lessons."
- Share books to use for read-alouds (in different genres). "My kids loved this today (this week or this month)."
- Share books to use for writing workshop mini-lessons. "Remember how I was talking about XXX. This book helped my mini-lesson so much."

- Review, integrate, and implement State Standards. "I'm stumped. Let's see if we can figure this out together."
- Gather resources for units of study. "I do/did a study like that. I'll show you some of the things I used."
- Team teach or co-teach a specific lesson. "How about if we try to teach it together?"

Tool for Practice: *Eight Steps to Getting Started on an Inquiry Project*

Step 1. Reflect upon and analyze your teaching and/or your students' learning or behaviors to identify a specific area of concern where there is a *difference* between the way things *are* and the way you *would like them to be.* For example, during your sustained silent reading time many of the students are jumping up and exchanging their books every five minutes. They don't seem to settle down to really engage with books. There is a lot of whispered chatter and tugging and pulling at each other instead of reading.

Step 2. Using the present tense, describe the situation as though everything was going well and the concern did not exist. What do you see, hear, smell, taste, feel in this ideal setting? What are others saying and doing? What are you saying and doing? For example, all the students have a book that they read during sustained silent reading. The books are stored in their desks and they take them out quickly, find their place with their bookmark, and relocate themselves to a favorite "spot" in the room to read. Within a few minutes a quiet calm takes over that is punctuated with natural rustlings as pages turn and bodies shift in chairs, on carpets, and in quiet spaces throughout the room. I say nothing to break their concentration. If a student gets up, we gesture to communicate. Silent reading is the powerful force and authority in the room.

Step 3. As specifically as you can, describe the concern you want to address. Write your thoughts in the form of a series of reflective questions and statements. Limit your focus to one particular concern. For example, I want to find out:

- Why are my students not engaged in reading for extended periods of time?
- Why do they keep getting up and exchanging their books?
- Why can't they seem to find a book that sustains their interest?

Step 4. Identify gaps in your abilities, knowledge, skills, understandings, and/or resources that may be keeping you from addressing your concern. For example,

- Am I establishing the condition they need to read without interruption?
- Am I supporting them in making appropriate book choices?

- Do I have enough books and an appropriate variety of books to meet their interests and levels?

Step 5. Identify potential resources that can inform your inquiry. For example, I need to see if there is an existing interest inventory online that I can use to find out what kinds of books my students like. I wonder if there is already a set of interview questions I could use. I'll see if there are any articles or readings that talk about how to engage students in reading for longer and longer periods of time so that I can find some good ideas to spark my own thinking.

Step 6. List specific actions you will undertake to fill or bridge the gaps you identified. Estimate a start and end date for each action. For example, next Friday I will conduct an interest inventory to determine the topics they like to read about. From Monday through Thursday, during silent reading, I'll observe the students to see if all the students are to read or if it is really just a few and those few are interfering with the others. By Friday of the following week, I will interview the book exchangers to determine why they aren't able to stick to a book and I'll interview some of the other students to find out what they would need in order to be able to read for a longer period of time.

Step 7. Commit to carrying out the activities. For example, don't make excuses, just do it!

Step 8. Assess the results. For example, I'll look at the inventory results, my observation notes, and the interview results to see what I can do to expand my library, support some students by creating the conditions they need to read, and help other students develop their book selection strategies.

Adapted from: http://teachers.net/gazette/DEC08/portner. Downloaded on December, 20, 2008, at 10:15 a.m. *Teachers Net Gazette* article by Hal Portner.

Tool for Practice: *Presentation Checklist*

- Did I identify a procedure, lesson, unit, or process that I designed that created a positive change in my classroom?
- Did I list everything I might want to say about what I did?
- Did I categorize my list into three–four major topics with at least two subtopics?
- Did I review my topics and subtopics with the following three questions in mind:
  1. What is my purpose for wanting to share this idea with others?
  2. What does my audience care about the most?
  3. What do I want the people to remember or be able to do when I am finished?

- To keep the audience's attention, interest and engagement, did I punctuate my topics and subtopics with:
  o real-life examples?
  o true stories?
  o relevant jokes?
  o hands-on activities?
- Did I prepare something for my audience to take home? Everyone likes to leave with something they feel they can use later.

Final thoughts:

- Practice a few times with colleagues or friends, or at home alone in front of a mirror.
- Train yourself to keep to the topic, use time wisely, and have everything you need readily accessible so that people aren't waiting for you to get set-up.
- Plan for disasters by anticipating what could go wrong and having a back-up plan. For example, what if the electricity goes out or the projector doesn't work, be sure to bring hand-outs to use instead (or give them away at the end of your presentation as a take-away).
- Always put your name, affiliation, date, and contact information on your presentation hand-outs.

Tool for Practice: *Are You Ready to Be a Teacher Leader? Self-Assessment Tool*

Read each of the following scenarios and label each of its following behavioral descriptors with a letter corresponding to how often you respond in that way:

N = I Never do this.

S = I Sometimes do this.

G = I Generally do this.

A = I Always do this.

1. At a faculty meeting, the principal announces that a new district initiative will require every teacher to devote at least thirty minutes once a week to health education in order to respond to the rising incidence of childhood obesity throughout the district. In response, you:
   a. Keep quiet, say nothing, and return to your classroom to conduct business as usual until you have no other choice.
   b. Catch the attention of a colleague across the table and roll your eyes.
   c. Commiserate with your neighbor about how difficult it will be to do this on top of the other demands on your time.
   d. Raise your hand and ask if there is an established curriculum and other relevant questions about implementation.

e. Suggest that you would be willing to work with a couple of other teachers to collect some resources and design a unit of study.

2. A colleague drops by and admires your students' work display in the hallway. She asks if you have any other ideas that she could use to show off what her students are doing. In response, you:
   a. Take offense because you think that she is calling you a show-off.
   b. Tell her it wasn't hard, you had the student teacher/paraprofessional do it.
   c. Talk to her for five minutes about how draining it is to have to put on a show for the principal but end up giving her nothing.
   d. Hand her a book on *Building Better Bulletin Boards.*
   e. Walk over to her classroom to see what her students had done, ask a few questions about the theme of their work, and then talk through some ideas without sounding like a "know-it-all."

3. After a lot of time and effort on your part, you perfected your own system for recording what goes on with each individual student during the writing workshop portion of your literacy instruction. The system includes documenting the status of their work each week, collecting evidence and annotating strengths and challenges for easy reference, and setting up a schedule for individual conferences that allows you to spend quality time with each student at least once every two weeks and is flexible enough to respond to "need-it-now" urgencies. When asked what you are doing during the writing workshop, you:
   a. Keep it to yourself.
   b. Wait and describe it to the principal at your performance evaluation.
   c. Share your colleagues' concerns about how difficult it is to stay on top of student work.
   d. Say that you are following some of the suggestions in Nancy Atwell's book.
   e. Describe what you have developed and offer a day and time for others to see it "in action" in case they want to adapt it for their own use.

Scoring:

- three or more G or A codes on responses numbered e: You are already on your way to being a teacher leader who shares and collaborates with others.
- three or more G or A codes on responses numbered d: You have made some steps toward becoming a teacher leader because you are willing to share ideas. But you only go so far and take little credit for what work you may have done on your own. What is holding you back?
- three or more G or A codes on responses numbered c: You communicate with colleagues by sharing their concerns but you don't take any steps toward being part of the solution. Try to break out of this pattern by asking only one question next time, "What can we do about it?"
- three or more G or A codes on responses numbered b: You don't appear to value your ideas or work enough to share them in a positive way with your colleagues. Hold back on the negativity next time and play "devil's

advocate" to what can go wrong. Force yourself to find a positive and express it openly.

- three or more G or A codes on responses numbered a: Hmm. You take your work very personally and tend to share nothing. Perhaps it's time to challenge yourself to slide open the doors of your work with others. Take a first step and ask for help on something that you know you need to improve.

Suggested Resources

http://www.teacherleaders.org/home. Virtual teacher leader network community. Is a good way to connect with other teacher leaders nationwide and to see how teacher leaders are impacting national policy and practice.

http://www.nsdc.org/index.cfm. National Staff Development Council, includes resources and publications for teacher leaders. Is a good outlet for teacher leadership work, presentations, and inquiry papers.

http://www.nbpts.org/. National Board of Professional Teaching Standards, the home page and organization for obtaining national board certification.

http://www.reading.org. The International Reading Association offers a teacher research grant that can fund an inquiry study. The guidelines for applying can be found on this link.

http://www.nwp.org/cs/public/print/programs/tic?x-t=projects.view. The National Writing Project has a teacher inquiry community that offers mini-grants for local projects of study in addition to memberships in their community and other supportive structures.

List of Selected Professional Associations

National Science Teachers Association (NSTA): http://www.nsta.org/.
National Council of Teachers of Mathematics (NCTM): http://www.nctm.org/.
National Council of Social Studies (NCSS): http://www.socialstudies.org/.
National Council of Teachers of English (NCTE): http://www.ncte.org/.
International Reading Association (IRA): http://www.reading.org/.
Association of Childhood Education International (ACEI): http://www.acei.org/.
National Association for the Education of Young Children (NAEYC): http://www.naeyc.org/.

References

Ackerman, R., & MacKenzie, S.B. (2006). Uncovering teacher leadership. *Educational Leadership, 63*(8), 66–70.

Barth, R. (2001). *Learning by heart.* San Francisco, CA: Jossey-Bass.

Harris, A. (2005). Teacher leadership: More than just a feel-good factor? *Leadership and Policy in Schools, 4,* 201–219.

Hatch, T., White, M.E., & Faigenbaum, D. (2005). Expertise, credibility and influence: How teachers can influence policy, advance research and improve performance. *Teachers College Record, 105,* 1004–1035.

Lieberman, A., & Miller, L. (2004). *Teacher leadership.* San Francisco, CA: Jossey-Bass.

Phelps, P.A. (2008). Helping teacher become leaders. *The Clearing House, 81*(3), 119–122.

Silva, D.Y., Gimbert, B., & Nolan, J. (2000). Sliding the doors: Locking and unlocking possibilities for teacher leadership. *Teachers College Record, 104*, 779–804.

Walpole, S., & Beauchat, K.A. (2008). Facilitating teacher study groups. Retrieved from www.literacycoachingonline.org/briefs/**StudyGroups**Brief.pdf. April 27, 2009.

Walpole, S., & Blamey, K.L. (2008). Elementary literacy coaches: The reality of dual roles. The Reading Teacher, 62(3), 222–231.

Index